PLANT BASED I
BEGINNNERS

GREENS FOR LIFE!

The Essential Plant Based Diet Meal Plan For Vegetarians and Vegans To Increase Metabolism While Reducing Inflammation and Live Longer

GRACE SOSA

Table of Contents

PART I 10

Vegetarian Cookbook 11

Chapter 1: Breakfast Recipes 13

Black Bean Bowl 14

Coconut Blueberry Ricotta Bowl 15

Broccoli Quiche 16

Tomato Bagel Sandwich 17

Cornmeal And Blueberry Pancakes 18

Breakfast Tortilla 20

Zucchini Frittata 21

Oatmeal And Strawberry Smoothie 22

Chapter 2: Appetizers Recipes 24

Buffalo Cauliflower 24

Garlic Bread And Veggie Delight 26

Spinach Parmesan Balls 28

Cheese Garlic Bread 29

Stuffed Mushrooms 30

Tomato Bruschetta 31

Spicy Pumpkin Seeds 32

Chapter 3: Soups & Side Dishes Recipes 33

Carrot Soup 33

Celery Soup 35

Tomato Soup And Halloumi Croutons 37

Baked Potatoes And Mushrooms With Spinach 39

Garlic Potatoes....40
Buttery Carrots....41
Chapter 4: Main Course Recipes....42
Nut And Tofu Loaf....42
Velvety Chickpea Curry....44
Tofu Pad Thai....46
Eggplant Parmesan....48
Veg Korma....49
Mac And Cheese....50
Sesame Noodles....51
Chapter 5: Dessert Recipes....52
Raspberry And Rosewater Sponge Cake....52
Easy Tiramisu....54
Chocolate Marquise....56
Lemon Syllabub....58
PART II....59
Vegan Cookbook....60
Chapter 1: Breakfast Recipes....63
Shamrock Sandwich....63
Breakfast Burrito....65
Gingerbread Waffles....67
Green Chickpeas And Toast....69
Asparagus And Tomato Quiche....70
Breakfast Bowl....72
Tofu Pancakes....74
Chapter 2: Side Dish Recipes....76
Baked Beans....76

Baked Potato Wedges 78
Black Beans And Quinoa 79
Roasted Lemon Garlic Broccoli 80
Spicy Tofu 81
Spanish Rice 82
Pepper And Lemon Pasta 83
Chapter 3: High Protein Recipes 84
Kale Salad With Spicy Tempeh Bits And Chickpeas 84
Protein Bars 87
Tofu And Spinach Scramble 89
Vegan Tacos 90
Grilled Tofu Steaks And Spinach Salad 92
Corn, Quinoa, And Edamame Salad 94
Lentil Soup 95
Chapter 4: Dessert Recipes 97
Chocolate Pudding 97
Orange Cake 99
Pumpkin Tofu Pie 101
Vegan Brownie 103
Vegan Cupcake 104
Vanilla Cake 106
Miracle Fudge 108
Chapter 5: Sauces & Dips Recipes 109
Tomato Jam 109
Walnut Kale Pesto 110
Ranch Dressing 111
PART III 112

Chapter 1: Identifying the Mediterranean Diet 113

Defining the Mediterranean Lifestyle 114

The Rules of the Mediterranean Diet 115

Chapter 2: Savory Mediterranean Meals 118

Mediterranean Feta Mac and Cheese 118

Chickpea Stew 118

Savory Mediterranean Breakfast Muffins 119

Mediterranean Breakfast Bake 120

Mediterranean Pastry Pinwheels 122

Chapter 3: Sweet Treats on the Mediterranean Diet 123

Greek Yogurt Parfait 123

Overnight Oats 124

Apple Whipped Yogurt 125

Chapter 4: Gourmet Meals on the Mediterranean Diet 126

Garlic-Roasted Salmon and Brussels Sprouts 126

Walnut Crusted Salmon with Rosemary 127

Spaghetti and Clams 128

Braised Lamb and Fennel 129

Mediterranean Cod 131

Baked Feta with Olive Tapenade 132

Chapter 5: 30-Minutes or Less Meals 133

Vegetarian Toss Together Mediterranean Pasta Salad 133

Vegetarian Aglio e Olio and Broccoli 134

Cilantro and Garlic Baked Salmon 135

Harissa Pasta 136

Chapter 6: 1-Hour-or-Less Meals 137

1 Hour Baked Cod 137
Grilled Chicken Mediterranean Salad 138
Lemon Herb Chicken and Potatoes One Pot Meal 140
Vegetarian Mediterranean Quiche 142
Herbed Lamb and Veggies 143
Chicken and Couscous Mediterranean Wraps 145
Sheet Pan Shrimp 146
Mediterranean Mahi Mahi 148
Chapter 7: Slow Cooker Meals 149
Slow Cooker Mediterranean Chicken 149
Slow Cooker Vegetarian Mediterranean Stew 151
Vegetarian Slow Cooker Quinoa 153
Slow-Cooked Chicken and Chickpea Soup 155
Slow Cooked Brisket 157
Vegan Bean Soup with Spinach 158
Moroccan Lentil Soup 159
Chapter 8: Vegetarian and Vegan Meals 161
Vegetarian Greek Stuffed Mushrooms 161
Vegetarian Cheesy Artichoke and Spinach Stuffed Squash 162
Vegan Mediterranean Buddha Bowl 164
Vegan Mediterranean Pasta 166
Vegetarian Zucchini Lasagna Rolls 168
Vegetarian Breakfast Sandwich 170
Vegan Breakfast Toast 172
Vegetarian Shakshouka 173

PART I

The vegetarian diet has gained immense popularity in the last few years. According to some studies, it has been found that an estimated 18% of the world population is vegetarian. Apart from all the environmental and ethical benefits of removing meat from the diet, a properly planned vegetarian diet can reduce the risk of various chronic diseases, improve diet quality, and help in losing weight. The vegetarian diet also does not include poultry and fish. The majority of the people opt for a vegetarian diet for personal or religious reasons and for ethical reasons, such as animal rights. Other people opt for it for various environmental reasons, like livestock production, which results in the emission of greenhouse gases.

A vegetarian diet comes along with a wide array of benefits. It has been found that vegetarians have a better quality of diet when compared to meat-eaters. They also have a higher intake of beneficial nutrients, such as vitamin C, fiber, magnesium, and vitamin E. Switching from a normal meat diet to a vegetarian diet can result to be an effective strategy in case you want to lose weight. For instance, in a review of twelve studies, it has been found that vegetarians can lose about four pounds of weight over eighteen weeks than non-vegetarians. Also, vegetarians have lower BMI or body mass index than non-vegetarians.

Some research found that a vegetarian diet can be linked to a lower risk of developing cancer and those of the colon, breast, stomach, and rectum. However, it lacks enough evidence to prove that a vegetarian diet can effectively reduce cancer risk. People who follow a vegetarian diet can maintain healthy levels of blood sugar. It can also help in preventing the onset of diabetes by controlling the levels of blood sugar. Vegetarian diets help in the reduction of various heart diseases that can make your heart stronger and healthier.

A vegetarian diet needs to include a wide mixture of veggies, fruits, healthy fats, grains, and proteins. To replace the protein that you get from meat in a diet, you will need to include plant foods rich in proteins such as seeds, nuts, legumes, seitan, tofu, and tempeh. Consuming whole foods rich in nutrients, such as vegetables and fruits, can provide your body with the necessary minerals and vitamins to fill the nutritional gaps in a diet. Some of the food items that you can include in your diet are:

- **Fruits:** Bananas, apples, melons, oranges, peaches, pears
- **Vegetables:** Asparagus, leafy greens, carrots, broccoli
- **Legumes:** Beans, lentils, chickpeas, peas
- **Seeds:** Chia, flaxseed, hemp
- **Nuts:** Walnuts, cashews, almonds
- **Proteins:** Seitan, natto, tofu, eggs, tempeh

You cannot include food items such as fish, seafood, meat, poultry, and ingredients that are based on meat. Restriction of eggs and dairy products are applicable for the vegans and not for vegetarians. I have included some tasty and easy vegetarian recipes that will help you to plan your diet effectively.

Chapter 1: Breakfast Recipes

No matter which diet you follow, breakfast is very important in all cases. In this section, you will find some easy-to-make and tasty breakfast recipes that you can include in your vegetarian diet.

Black Bean Bowl

Total Prep & Cooking Time: Fifteen minutes

Yields: Two servings

Nutrition Facts: Calories: 620.1 | Protein: 28.9g | Carbs: 47.6g | Fat: 37.1g | Fiber: 23.2g

Ingredients:

- Two tbsps. of olive oil
- Four eggs (beaten)
- One can of black beans (rinsed)
- One avocado (sliced)
- One-fourth cup of salsa
- One tsp. of each
    - Black pepper (ground)
    - Salt

Method:

1. Take a small pan and heat oil in it. Add the eggs and scramble for five minutes.

2. Place the beans in a bowl. Heat the beans in the oven for one minute.

3. Divide the beans into two serving bowls.

4. Top the beans with scrambled eggs, salsa, and avocado. Add pepper and salt according to taste.

Coconut Blueberry Ricotta Bowl

Total Prep & Cooking Time: Twenty minutes

Yields: One serving

Nutrition Facts: Calories: 303.2 | Protein: 9.5g | Carbs: 32.1g | Fat: 15.3g | Fiber: 3.9g

Ingredients:

- One-fourth cup of ricotta cheese
- One tbsp. of each
  - Honey
  - Coconut milk
  - Slivered almonds
  - Coconut flakes
- Half cup of blueberries

Method:

1. Combine coconut milk and ricotta in a medium-sized bowl. Drizzle honey from the top and add coconut and almonds.

2. Serve with blueberries from the top.

Broccoli Quiche

Total Prep & Cooking Time: Fifty minutes

Yields: Six servings

Nutrition Facts: Calories: 378.2 | Protein: 17.3g | Carbs: 21.1g | Fat: 25.8g | Fiber: 3.1g

Ingredients:

- One pie crust (unbaked)
- Three tbsps. of butter
- One onion (minced)
- One tsp. of each
    - Garlic (minced)
    - Salt
- Two cups of broccoli (chopped)
- One and a half cup of mozzarella cheese (shredded)
- Four eggs (beaten)
- Two and a half cup of milk
- Half tsp. of black pepper
- One tbsp. of butter (melted)

Method

1. Start by preheating your oven at 175 degrees Celsius. Use pie crust for lining a deep pie pan.

2. Take a large saucepan and add butter to it. Add broccoli, onion, and garlic. Cook the veggies slowly until tender. Add the cooked veggies to the pie crust and add cheese from the top.

3. Mix milk and eggs in a bowl; add pepper and salt for seasoning. Add the remaining butter to the milk mixture. Pour the mixture over the mixture of vegetables.

4. Bake the quiche in the oven for forty minutes or until the center has properly set

Tomato Bagel Sandwich

Total Prep & Cooking Time: Twenty minutes

Yields: One serving

Nutrition Facts: Calories: 346.3 | Protein: 13.1g | Carbs: 48.6g | Fat: 10.7g | Fiber: 2.9g

Ingredients:

- One bagel (split, toasted)
- Two tbsps. of cream cheese
- One large tomato (sliced thinly)
- Pepper and salt (to season)
- Four basil leaves

Method:

1. Spread the cream cheese on the halves of the bagel.
2. Top the cheese layer with slices of tomato. Add pepper and salt for seasoning.
3. Serve with basil leaves from the top.

Cornmeal And Blueberry Pancakes

Total Prep & Cooking Time: Thirty minutes

Yields: Six servings

Nutrition Facts: Calories: 180.3 | Protein: 5.2g | Carbs: 26.8g | Fat: 5.1g | Fiber: 4.6g

Ingredients:

- One cup of soy milk
- Half cup of each
    - Water
    - Cornmeal (ground)
- One and a half cup of wheat flour
- One tsp. baking powder
- One-third tsp. baking soda
- One-fourth tsp. of salt
- One cup of blueberries
- Two tbsps. of vegetable oil

Method:

1. Preheat your oven at 95 degrees Celsius.

2. Combine water and soy milk in a bowl.

3. Take a large mixing bowl and combine cornmeal, baking soda, flour, salt, and baking powder. Add the mixture of soy milk. Combine properly. Add the blueberries and allow the batter to rest for five minutes.

4. Take a large skillet and grease it using oil. Add one-fourth cup of the batter in the skillet. Cook until the pancakes are bubbly on the top, and the edges are dry. Cook for five minutes on each side. Repeat with the remaining batter.

5. Serve hot with jam or syrup.

Breakfast Tortilla

Total Prep & Cooking Time: Twenty minutes

Yields: Two servings

Nutrition Facts: Calories: 381.3 | Protein: 15.8g | Carbs: 38.1g | Fat: 18.7g | Fiber: 4.5g

Ingredients:

- Two tbsps. of beans (refried)
- Three tbsps. of salsa
- Three large eggs (beaten)
- One tbsp. of mayonnaise
- Four tortillas (flour)
- One and a half cup of lettuce (shredded)

Method:

1. Combine salsa and beans in a small bowl.

2. Take an iron skillet and heat oil in it. Add the eggs and let the bottom set—Cook for one minute. Spread the mixture of beans on half of the egg and flip one side for making the shape of a half-circle. Cook until the eggs set properly.

3. Spread mayonnaise on the tortillas.

4. Cut the cooked eggs into four equal pieces. Place each piece of eggs on the tortillas. Top with lettuce.

5. Roll the tortillas. Serve hot.

Zucchini Frittata

Total Prep & Cooking Time: Forty minutes

Yields: Five servings

Nutrition Facts: Calories: 258.3 | Protein: 14.2g | Carbs: 9.1g | Fat: 19.6g | Fiber: 3.4g

Ingredients:

- One cup of water
- Three tbsps. of olive oil
- Half tsp. of salt
- Half bell pepper (green, chopped)
- Three zucchinis (cut in slices of half-inch)
- Two garlic cloves (peeled)
- One onion (diced)
- Six mushrooms (chopped)
- One tbsp. of butter
- Five eggs
- Pepper and salt (to taste)
- One and a half cup of mozzarella cheese (shredded)
- Three tbsps. of parmesan cheese

Method:

1. Start by preheating your oven at 160/175 degrees Celsius.

2. Take a large skillet and combine olive oil, water, green pepper, salt, garlic cloves, and zucchini. Simmer the mixture until the zucchini is soft—Cook for seven minutes.

3. Drain the water and remove the garlic; add mushroom, onion, and butter. Keep cooking until the onion turns transparent. Add the eggs and keep stirring. Add pepper and salt for seasoning. Cook until the eggs are firm.

4. Add mozzarella cheese from the top.

5. Bake in the oven for ten minutes.

6. Remove the frittata from the oven and add parmesan cheese from the top. Place under the broiler for about five minutes.

7. Cut the frittata in wedges and serve warm.

Oatmeal And Strawberry Smoothie

Total Prep & Cooking Time: Twenty minutes

Yields: Two servings

Nutrition Facts: Calories: 204.1 | Protein: 5.2g | Carbs: 41.3g | Fat: 2.7g | Fiber: 6.9g

Ingredients:

- One cup of almond milk
- Half cup of rolled oats
- Fourteen strawberries (frozen)
- One banana (cut in chunks)
- Two tsps. of agave nectar
- Half tsp. of vanilla extract

Method:

1. Add almond milk, strawberries, oats, agave nectar, banana, and vanilla extract in a food processor. Keep blending until smooth.
2. Serve with pieces of strawberry from the top.

Chapter 2: Appetizers Recipes

Hosting a party but not sure which appetizers to prepare as most of your guests are vegetarians? No worries as I have included some great vegetarian recipes for appetizers in this section.

Buffalo Cauliflower

Total Prep & Cooking Time: Thirty minutes

Yields: Twelve servings

Nutrition Facts: Calories: 120 | Protein: 4.2g | Carbs: 7.9g | Fat: 8.7g | Fiber: 2.3g

Ingredients:

- One serving of cooking spray
- Half cup of buffalo sauce
- Three tbsps. of mayonnaise
- One large egg
- Six cups of cauliflower florets
- Two cups of garlic croutons
- One-fourth cup of parmesan cheese (grated)

For the dipping sauce:

- One-fourth cup of each
    - Sour cream
    - Blue cheese salad dressing
- One tsp. of black pepper (ground)

Method:

1. Start by preheating your oven to 230 degrees Celsius. Use a cooking spray for greasing a baking tray.

2. Combine mayonnaise, buffalo sauce, and egg in a bowl. Toss the florets of cauliflower in the mixture of sauce and coat properly.

3. Spread the tossed florets on the baking tray.

4. Add the croutons on a blender and pulse them into crumbs. Add the cheese and pulse again. Spread the mixture of cheese and croutons over the florets of cauliflower.

5. Bake for fifteen minutes until tender and browned. Allow the florets to sit for five minutes.

6. Mix all the ingredients for the dipping sauce.

7. Serve cauliflower florets with dip sauce by the side.

Garlic Bread And Veggie Delight

Total Prep & Cooking Time: Thirty minutes

Yields: Five servings

Nutrition Facts: Calories: 390.2 | Protein: 12.4g | Carbs: 58.7g | Fat: 11.6g | Fiber: 7.2g

Ingredients:

- One cup of olive oil
- One garlic clove (chopped)
- One eggplant (cubed)
- One zucchini (cubed)
- One tomato (chopped)
- One tsp. of salt
- Two tsps. of each
  - Basil (minced)
  - Oregano (minced)
- One baguette
- Four tsps. of garlic powder
- Six tsps. of butter

Method:

1. Take a large skillet and add olive oil to it. Add garlic and fry for two minutes until browned.
2. Add the zucchini and eggplant to the skillet and cook for five minutes. Make sure that the eggplant is tender and brown.
3. Add the chunks of tomato and combine the veggies; add basil, oregano, and salt. Cook for two minutes and remove from heat.
4. Preheat oven 140/165 degrees Celsius.
5. Slice the baguette into one-inch slices, approximately twelve slices. Add butter and garlic powder on the bread slices and place them on the oven rack. Heat the bread for five minutes.
6. Arrange the heated bread slices on a plate. Top the slices with vegetables.
7. Serve immediately.

Spinach Parmesan Balls

Total Prep & Cooking Time: Thirty minutes

Yields: Ten servings

Nutrition Facts: Calories: 254.1 | Protein: 11.4g | Carbs: 18.4g | Fat: 14.2g | Fiber: 3.5g

Ingredients:

- Twenty ounces of frozen spinach (chopped)
- Two cups of bread crumbs
- One cup of parmesan cheese (grated)
- Half cup of butter (melted)
- Four green onions (chopped)
- Four eggs (beaten)
- Pepper and salt (for seasoning)

Method:

1. Start by preheating your oven at 175 degrees Celsius.
2. Take a bowl and combine spinach, bread crumbs, cheese, green onions, butter, pepper, salt, and eggs. Make balls of one-inch size from the prepared mixture.
3. Arrange the spinach balls on a baking tray. Bake for fifteen minutes until browned.
4. Serve hot.

Cheese Garlic Bread

Total Prep & Cooking Time: Thirty minutes

Yields: Eight servings

Nutrition Facts: Calories: 260 | Protein: 7.1g | Carbs: 29.7g | Fat: 11.4g | Fiber: 1.5g

Ingredients:

- Half cup of butter (melted)
- One tsp. of garlic salt
- One-fourth tsp. of rosemary (dried)
- One-eighth tsp. of each
  - Basil (dried)
  - Thyme (dried)
  - Garlic powder
- One tbsp. of parmesan cheese (grated)
- One loaf of French bread (halved)

Method:

1. Preheat oven at 150 degrees Celsius.
2. Mix garlic salt, butter, basil, rosemary, thyme, garlic powder, and cheese in a bowl.
3. Spread the butter mixture on the halves of the bread. Add extra cheese from the top if you want to.
4. Place halves of bread on a baking tray. Bake for twelve minutes until browned.

Stuffed Mushrooms

Total Prep & Cooking Time: Forty-five minutes

Yields: Twelve servings

Nutrition Facts: Calories: 89 | Protein: 2.6g | Carbs: 1.3g | Fat: 8.9g | Fiber: 0.6g

Ingredients:

- Twelve whole mushrooms
- One tbsp. of each
  - Minced garlic
  - Vegetable oil
- Eight ounces of cream cheese (softened)
- One-fourth cup of parmesan cheese (grated)
- One-fourth tsp. of each
  - Onion powder
  - Black pepper (ground)
  - Cayenne powder (ground)

Method:

1. Preheat the oven to 175 degrees Celsius. Grease a baking tray with the help of cooking spray.

2. Clean the mushrooms using a damp kitchen towel; break the stems. Chop the mushroom stems finely.

3. Take a skillet and heat oil in it. Add chopped stems of mushroom and garlic—Cook for five minutes.

4. Remove the skillet from heat and let it cool. Add the cream cheese, black pepper, parmesan cheese, cayenne powder, and onion powder. Mix well.

5. Use a small spoon for filling the mushroom caps with the mushroom stuffing.

6. Place the mushroom caps on the prepared baking tray.

7. Bake for twenty minutes until liquid forms under the mushroom caps.

Tomato Bruschetta

Total Prep & Cooking Time: Thirty-five minutes

Yields: Twelve servings

Nutrition Facts: Calories: 214.1 | Protein: 10.6g | Carbs: 23.1g | Fat: 8.8g | Fiber: 2.6g

Ingredients:

- Six tomatoes (chopped)
- Half cup of sun-dried tomatoes
- Three garlic cloves (minced)
- One-fourth cup of olive oil
- Two tbsps. of balsamic vinegar
- One-third cup of basil
- One-fourth tsp. of each
    - Black pepper (ground)
    - Salt
- One baguette
- Two cups of mozzarella cheese (shredded)

Method:

1. Preheat your oven on the setting of broiler.
2. Combine tomatoes, olive oil, vinegar, garlic, sun-dried tomatoes, basil, pepper, and salt in a bowl. Let the mixture sit for ten minutes.
3. Cut the baguette into slices of a three-fourth inch. Arrange the baguette slices on a baking tray. Broil for two minutes until browned.
4. Add the mixture of tomatoes on the slices of bread and top with mozzarella cheese.
5. Broil again for five minutes.

Spicy Pumpkin Seeds

Total Prep & Cooking Time: One hour and ten minutes

Yields: Eight servings

Nutrition Facts: Calories: 91 | Protein: 3.2g | Carbs: 8.8g | Fat: 4.9g | Fiber: 0.7g

Ingredients:

- Two tbsps. of margarine
- Half tsp. of salt
- One-eighth tsp. of garlic salt
- Two tsps. of Worcestershire sauce
- Two cups of pumpkin seeds (raw)

Method:

1. Preheat your oven at 135 degrees Celsius.
2. Combine the ingredients in a mixing bowl.
3. Bake for one hour. Stir in between.

Chapter 3: Soups & Side Dishes Recipes

Soup forms an integral part of a vegetarian diet along with the side dishes. Here are some easy vegetarian recipes for side dishes and soups that you can include in your diet.

Carrot Soup

Total Prep & Cooking Time: Forty-five minutes

Yields: Four servings

Nutrition Facts: Calories: 351 | Protein: 3.2g | Carbs: 23.1g | Fat: 29.7g | Fiber: 4.2g

Ingredients:

- Two tbsps. of olive oil
- Four-hundred grams of carrots (cut in disks of half-inch)
- Half onion (diced)
- Four cloves of garlic (smashed)
- Two tsps. of cumin seeds
- Four cups of vegetable stock
- Two bay leaves
- One tsp. of salt
- One-fourth tsp. of white pepper
- Two tsps. of honey
- One-fourth cup of yogurt

Method:

1. Take a heavy bottom pan and add oil in it. Add onions, garlic, and cumin to the pan. Sauté on a medium flame for six minutes until tender and golden in color. Stir occasionally.

2. Add the stock, carrots, salt, bay leaves, white pepper, and simmer the mixture. Cover the pan and simmer for twenty minutes. Allow the soup to cool down for five minutes.

3. Use an immersion blender for blending the soup. Blend until you reach a silky smooth consistency.

4. Return the soup to heat and add honey. Stir the soup. Add yogurt and simmer. Taste the soup and adjust the seasonings. Keep the soup warm on very low flame until you serve.

5. Divide the soup among serving bowls. Serve with a dollop of yogurt from the top.

Note: You can use ground spices in place of the whole spices. But, whole spices will help in adding extra flavor.

Celery Soup

Total Prep & Cooking Time: Thirty-five minutes

Yields: Seven servings

Nutrition Facts: Calories: 180 | Protein: 4.2g | Carbs: 23.1g | Fat: 9.2g | Fiber: 4.9g

Ingredients:

- Two tbsps. of olive oil
- One onion (diced)
- Four cloves of garlic (chopped)
- Six cups of celery (thinly sliced)
- Two cups of potatoes (sliced in rounds)
- Four cups of vegetable stock

- One cup of water
- One bay leaf
- One tsp. of salt
- Half tsp. of pepper
- One-third tsp. of cayenne
- Half cup of sour cream
- One-fourth cup of each
    - Parsley (small stems)
    - Dill (small stems)

Method:

1. Take a large pot and add oil in it. Heat the oil and start adding the onion. Cook for five minutes until golden.

2. Roughly chop celery, potatoes, and garlic. Add garlic and cook the mixture for two minutes. Add potatoes, celery, stock, bay leaf, water, salt, cayenne, and pepper. The liquid needs to be enough to cover the vegetables. Cover the pot and boil the mixture. Simmer for ten minutes.

3. Remove bay leaf after turning off the stove. Add herbs to the pot and allow them to wilt.

4. Take an immersion blender and start blending the soup until silky smooth.

5. Return the pot to heat and cook over low flame for five minutes.

6. Serve with sour cream from the top.

Tomato Soup And Halloumi Croutons

Total Prep & Cooking Time: One hour and five minutes

Yields: Six servings

Nutrition Facts: Calories: 285 | Protein: 10.3g | Carbs: 13.2g | Fat: 23.2g | Fiber: 4.9g

Ingredients:

- Three pounds of tomatoes
- Half red onion (sliced in thin rings)
- Six cloves of garlic
- Two tsps. of thyme leaves

- One-third cup of oil
- Four cups of vegetable stock
- One-fourth cup of basil leaves (chopped)
- One cup of Greek yogurt

For croutons:

- One block of halloumi cheese (cut in cubes of three-fourth inch)
- One tbsp. of oil

Method:

1. Start by preheating your oven at 200 degrees Celsius.
2. Use parchment paper for lining baking sheet. Spread the onions, tomatoes, and garlic on the sheet. Drizzle with some oil from the top—roast in the oven for thirty minutes.
3. Heat some oil in a pan and start adding the halloumi cubes. Cook for four minutes until golden on all sides.
4. Add the roasted veggies in a pot along with the vegetable stock. Use an immersion blender for blending the soup until smooth. Place the pot over a low flame and add seasonings of your choice. Simmer the soup and add basil. Simmer for ten minutes.
5. Add half a cup of yogurt to the soup.
6. Serve the soup in serving bowls with croutons from the top.

Baked Potatoes And Mushrooms With Spinach

Total Prep & Cooking Time: Forty-five minutes

Yields: Four servings

Nutrition Facts: Calories: 235.1 | Protein: 7.1g | Carbs: 27.1g | Fat: 11.3g | Fiber: 4.9g

Ingredients:

- One pound of potatoes (halved)
- Three tbsps. of olive oil
- Half pound of Portobello mushroom
- Six garlic cloves
- Two tbsps. of thyme (chopped)
- One pinch of black pepper and salt
- One-fourth cup of cherry tomatoes
- Half cup of spinach (sliced)
- Two tbsps. of pine nuts (toasted)

Method:

1. Preheat your oven at 200/220 degrees Celsius.
2. Add the potatoes in a roasting pan and drizzle some oil from the top. Roast the potatoes for fifteen minutes.
3. Add mushrooms along with garlic to the pan. Add thyme from the top along with some olive oil. Sprinkle black pepper and salt. Roast again for five minutes.
4. Add cherry tomatoes to the pan. Cook again for five minutes until the mushrooms are tender.
5. Add toasted pine nuts from the top and serve with spinach by the side.

Garlic Potatoes

Total Prep & Cooking Time: Fifty minutes

Yields: Four servings

Nutrition Facts: Calories: 270.2 | Protein: 5.2g | Carbs: 39.7g | Fat: 12.1g | Fiber: 4.9g

Ingredients:

- Two pounds of red potatoes (quartered)
- One-fourth cup of butter (melted)
- Two tsps. of garlic (minced)
- One tsp. of salt
- One lemon (juiced)
- One tbsp. of parmesan cheese (grated)

Method:

1. Start by preheating the oven at 175 degrees Celsius.
2. Place the potatoes in a baking dish.
3. Combine butter, lemon juice, garlic, and salt in a small bowl. Add this mixture over the potatoes and stir for coating. Add parmesan cheese from the top.

4. Bake the potatoes by covering the dish in the preheated oven for thirty minutes. Remove the cover and bake again for ten minutes.

Buttery Carrots

Total Prep & Cooking Time: Twenty-five minutes

Yields: Four servings

Nutrition Facts: Calories: 182.3 | Protein: 0.8g | Carbs: 21.3g | Fat: 10.3g | Fiber: 3.9g

Ingredients:

- One pound of baby carrots
- One-fourth cup of margarine
- One-third cup of brown sugar

Method:

1. Cook the baby carrots in boiling water. Drain most of the liquid leaving behind a little bit of liquid at the base.

2. Remove the carrots from the pot. Add brown sugar along with margarine. Simmer the mixture for two minutes and add the carrots to the pot. Toss well for combining.

3. Serve warm.

Chapter 4: Main Course Recipes

After you are done with the soups and side dishes, now it is time to jump into the main course. Here are some tasty main course recipes that you can include within your vegetarian diet plan.

Nut And Tofu Loaf

Total Prep & Cooking Time: One hour and forty minutes

Yields: Six servings

Nutrition Facts: Calories: 308.3 | Protein: 15.2g | Carbs: 27.6g | Fat: 14.2g | Fiber: 4.8g

Ingredients:

- One serving of cooking spray
- Twelve ounces of tofu (firm, drained, cubed)
- Two large eggs
- One ounce dry mix of onion soup
- One tbsp. of soy sauce
- Three-fourth cup of walnuts (chopped)
- One tsp. of olive oil
- Eight ounces of fresh mushrooms (sliced)
- One onion (chopped)
- Two celery stalks (chopped)
- Two tsps. of oregano (dried)
- One and a half tsps. of basil (dried)
- Half tsp. of sage (dried)
- Two cups of bread crumbs

Method:

1. Start by preheating the oven at 175 degrees Celsius. Use a cooking spray for greasing loaf pan.

2. Place eggs, tofu, soy sauce, and onion soup mix in a blender. Blend the ingredients until properly combined. Add the walnuts and blend again. Transfer the mixture of tofu to a bowl.

3. Take a large skillet and heat oil in it. Add mushrooms and cook them for four minutes. Add celery and cook for two minutes—season with basil, sage, and oregano.

4. Stir bread crumbs and veggies into the mixture of tofu. Press the loaf mixture into the pan.

5. Bake the loaf in the oven for sixty to seventy minutes.

6. Let the loaf cool down for five minutes before serving.

7. Slice the loaf and serve warm.

Velvety Chickpea Curry

Total Prep & Cooking Time: Forty-five minutes

Yields: Six servings

Nutrition Facts: Calories: 408.3 | Protein: 10.2g | Carbs: 68.3g | Fat: 11.2g | Fiber: 8.9g

Ingredients:

- One tbsp. of each
    - Ginger root (minced)
    - Coconut oil
- One onion (sliced)
- Four garlic cloves (minced)
- Two tbsps. of curry powder
- One-fourth tsp. of pepper flakes
- Three cups of vegetable stock
- Two tbsps. of each
    - Soy sauce
    - Tomato paste
    - Maple syrup
- Half pound of potatoes (cut in pieces of a three-fourth inch)
- One carrot (sliced)
- Four cups of cauliflower florets
- One can of chickpeas (rinsed)
- Half cup of coconut milk
- One-fourth cup of cilantro (chopped)
- Half cup of peas (frozen)
- Salt (for seasoning)

Method:

1. Take a heavy-based pot and melt coconut oil in it. Add onions to the pot and sauté for five minutes. Add garlic and ginger to the pot—Cook for thirty seconds. Add pepper flakes, curry powder, soy sauce, stock, tomato paste, and maple syrup. Stir well.

2. Add carrots and potatoes to the pot and cover. Boil the mixture. Slightly open the cover and simmer for ten minutes. Add chickpeas, cauliflower, cilantro, and coconut milk. Stir well for combining. Simmer again for seven minutes. Add peas and cook for one minute.

3. Season with salt and cook for one minute.

4. Serve with basmati rice and cilantro from the top.

Tofu Pad Thai

Total Prep & Cooking Time: Forty-five minutes

Yields: Four servings

Nutrition Facts: Calories: 451 | Protein: 15.4g | Carbs: 60.1g | Fat: 15.2g | Fiber: 4.2g

Ingredients:

- Twelve ounces of tofu (drained, cubed)
- One tbsp. of cornstarch
- Three tbsps. of vegetable oil
- Eight ounces of rice noodles

For the sauce:

- One-fourth cup of each
    - Sriracha sauce
    - Water
    - Soy sauce
- Two tbsps. of white sugar
- One tbsp. of tamarind concentrate
- One tsp. of pepper flakes
- One large egg

- Two tbsps. of spring onions (chopped)
- One and a half tbsp. of peanuts (crushed)
- One lime (cut in wedges)

Method:

1. Coat the tofu cubes with cornstarch in a large bowl.
2. Heat two tbsps. of oil in a skillet. Fry the coated tofu for two minutes on each side.
3. Place the rice noodles in a medium bowl. Cover the noodles using hot boiling water. Soak the noodles until soft for three minutes. Drain the water.
4. Mix sriracha sauce, water, sugar, soy sauce, pepper flakes, and tamarind concentrate in a skillet. Cook for five minutes.
5. Heat one tbsp. of oil in a large wok. Add noodles, onion, along with the tofu. Cook the mixture for three minutes. Add sauce and toss it for combining.
6. Push the noodles to a side and crack the egg in the center. Stir for thirty seconds and mix with the noodles.
7. Serve with peanuts, green onion, and wedges of lime.

Eggplant Parmesan

Total Prep & Cooking Time: One hour

Yields: Ten servings

Nutrition Facts: Calories: 480.2 | Protein: 21.2g | Carbs: 60.1g | Fat: 15.2g | Fiber: 9.8g

Ingredients:

- Three eggplants (sliced)
- Two eggs (beaten)
- Four cups of bread crumbs
- Six cups of spaghetti sauce
- Sixteen ounces of mozzarella cheese (shredded)
- Half cup of parmesan cheese
- Half tsp. of basil (dried)

Method:

1. Preheat the oven at 175 degrees Celsius.
2. Dip the slices of eggplant in egg and then coat in bread crumbs.
3. Arrange the slices of eggplant in a baking sheet and bake for five minutes on both sides.
4. Take a baking dish and spread the spaghetti sauce for covering the base. Arrange the eggplant slices over the sauce. Sprinkle cheese from the top. Repeat for the remaining layers. Top with basil and cheese.
5. Bake for thirty-minutes until browned.

Veg Korma

Total Prep & Cooking Time: Fifty-five minutes

Yields: Four servings

Nutrition Facts: Calories: 451 | Protein: 8.4g | Carbs: 40.1g | Fat: 30.2g | Fiber: 8.7g

Ingredients:

- Two tbsps. of vegetable oil
- One onion (diced)
- One tsp. of ginger root (minced)
- Four garlic cloves (minced)
- Two potatoes (cubed)
- Four carrots (cubed)
- One jalapeno pepper (sliced)
- Three tbsps. of cashews (ground)
- One can of tomato sauce
- Two tsps. of salt
- One and a half tbsps. of curry powder
- One cup of green peas (frozen)
- One red bell pepper (roughly chopped)
- One-third yellow bell pepper (roughly chopped)
- One bunch of cilantro
- Half cup of heavy cream

Method:

1. Take a skillet and heat oil in it. Add onions to the oil and cook until soft. Add garlic and ginger to the skillet. Cook for one minute. Add carrots, potatoes, cashews, jalapenos, and tomato sauce. Add curry powder and season with salt. Stir well and cook for ten minutes until the potatoes are soft.

2. Add bell pepper, peas, and cream. Lower the flame and cover the skillet—Cook for ten minutes.

3. Serve with cilantro from the top.

Mac And Cheese

Total Prep & Cooking Time: Fifty minutes

Yields: Six servings

Nutrition Facts: Calories: 456 | Protein: 24g | Carbs: 33.1g | Fat: 24.9g | Fiber: 2.3g

Ingredients:

- Two cups of elbow macaroni (uncooked)
- One-fourth cup of butter
- Two tbsps. of flour
- One tsp. of each
    - Black pepper (ground)
    - Mustard powder
- Two cups of milk
- Eight ounces of each
    - Cheese food (cubed)
    - American cheese (cubed)
- Half cup of bread crumbs

Method:

1. Preheat the oven at 180/200 degrees Celsius. Take a casserole dish and grease with butter.

2. Boil water in a pot with salt. Add the pasta and cook the pasta for six minutes. Drain the water.

3. Take a saucepan. Melt butter in it. Add mustard powder, flour, and pepper. Add milk and stir constantly. Add the cheeses and mix for two minutes until the sauce thickens.

4. Add the macaroni to the cheese mixture. Mix well.

5. Transfer the pasta mixture to the greased dish. Add bread crumbs from the top.

6. Bake for twenty minutes by covering the dish.

7. Serve hot.

Sesame Noodles

Total Prep & Cooking Time: Thirty minutes

Yields: Eight servings

Nutrition Facts: Calories: 365.2 | Protein: 7.3g | Carbs: 51.1g | Fat: 13.2g | Fiber: 3.9g

Ingredients:

- Sixteen ounces of linguine pasta
- Six garlic cloves (minced)
- Six tbsps. of each
    - Safflower oil
    - Sugar
    - Rice vinegar
    - Soy sauce
- Two tsps. of chili sauce
- Two tbsps. of sesame oil
- Six green onions (sliced)
- One tsp. of sesame seeds (toasted)

Method:

1. Boil water in a large pot along with some salt. Add the pasta. Keep cooking for eight minutes. Drain all the water.

2. Take a saucepan and heat oil in it. Add garlic, sugar, soy sauce, chili sauce, and sesame oil. Boil the mixture until the sugar gets dissolved.

3. Add the sauce to the cooked pasta and toss well for combining.

4. Serve with sesame seeds and green onions from the top.

Chapter 5: Dessert Recipes

Having a great dessert after a tasty meal can make you, as well as your stomach, feel good. So, here are some vegetarian dessert recipes for you.

Raspberry And Rosewater Sponge Cake

Total Prep & Cooking Time: Fifty-five minutes

Yields: Ten servings

Nutrition Facts: Calories: 441 | Protein: 4.1g | Carbs: 51.3g | Fat: 23.1g | Fiber: 1.3g

Ingredients:

- Two-hundred grams of butter (softened)
- Two-hundred and fifty grams of caster sugar
- Four eggs (beaten)
- One tsp. of vanilla extract
- Two cups of flour

For the rose filling:

- Half cup of double cream
- One tsp. of rosewater
- Four tbsps. of raspberry jam
- One-third cup of raspberries (crushed)

For rose icing:

- Half cup of icing sugar
- Half tsp. of rosewater

Method:

1. Heat the oven at 160 degrees Celsius.
2. Use parchment paper for lining two baking tins—grease with butter.
3. Mix sugar and butter in a bowl. Add the eggs and mix again.
4. Add vanilla extract to the egg mixture and mix. Add flour and fold.
5. Divide the cake batter into the prepared baking tins and bake in the oven for twenty minutes. Let the cakes cool for ten minutes.
6. Whisk double cream along with rosewater. Add the jam and mix.
7. Place one of the cakes on a serving plate and add the cream mixture. Add raspberries from the top and place the other cake on top.
8. Mix all the ingredients for the icing.
9. Serve the cake with rose icing from the top.

Easy Tiramisu

Total Prep & Cooking Time: One hour and fifteen minutes

Yields: Two servings

Nutrition Facts: Calories: 741 | Protein: 11.3g | Carbs: 44.3g | Fat: 51.6g | Fiber: 1.9g

Ingredients:

- Three tsps. of coffee granules
- Three tbsps. of coffee liqueur
- One and a half cup of mascarpone
- Half cup of condensed milk
- Six sponge fingers
- One tbsp. of cocoa powder

Method:

1. Mix coffee granules in two tbsps. of boiling water and stir for combining. Add three tbsps. of cold water along with coffee liqueur. Pour the mixture in a dish and keep aside.

2. Beat condensed milk, mascarpone, and vanilla extract in a bowl using a hand blender.

3. Break the fingers into two pieces and soak them in the mixture of coffee for thirty seconds.

4. Take a sundae glass and add the sponge fingers at the base. Add the cream mixture on top. Sift cocoa powder and chill in the refrigerator for one hour.

Chocolate Marquise

Total Prep & Cooking Time: Two hours and fifty-five minutes

Yields: Ten servings

Nutrition Facts: Calories: 710.3 | Protein: 7.8g | Carbs: 59.8g | Fat: 53g | Fiber: 1.5g

Ingredients:

- Three-hundred grams of dark chocolate
- Half cup of each
    - Caster sugar
    - Butter (softened)
- Six tbsps. of cocoa powder
- Six eggs
- One-third cup of double cream
- One-fourth cup of mint chocolate

Method:

1. Break the dark chocolate in small pieces and melt it using a double boiler system.
2. Beat butter and sugar in a bowl until creamy.

3. Separate the egg whites and yolks.

4. Mix the yolks with the sugar mixture until creamy.

5. Whip the double cream in another bowl.

6. Add the melted chocolate in the mixture of butter and fold gently. Add the whipped cream and mix well.

7. Spoon the mixture of chocolate in a piping bag.

8. Take a baking tin and pipe one layer of chocolate at the base of the tin. Cover the tin with pieces of mint chocolate. Repeat for the other layers. You will need to make four layers of mint chocolate.

9. Cover the tin with a cling film.

10. Chill the marquise in the fridge for two hours.

11. Remove the marquise from the tin by using a sharp knife.

12. Serve by cutting into slices.

Lemon Syllabub

Total Prep & Cooking Time: Fifteen minutes

Yields: Four servings

Nutrition Facts: Calories: 328 | Protein: 2.2g | Carbs: 14.6g | Fat: 28.6g | Fiber: 0.2g

Ingredients:

- Two cups of whipping cream
- Half cup of caster sugar
- Three tbsps. of white wine
- Half a lemon (juice and zest)
- Berries (for serving)

Method:

1. Mix sugar and whipping cream in a bowl. Whip until soft peaks are formed.
2. Add white wine in the mixture. Mix well. Add lemon juice and lemon zest in the mixture. Combine the ingredients properly.
3. Spoon the mixture into serving bowls or glasses.
4. Sprinkle remaining lemon zest from the top.
5. Serve the lemon syllabub with berries.

Note: You can use a mix of berries or only one type of berry

PART II

Vegan Cookbook

The vegan diet has gained immense popularity in the past few years. With an increasing number of participants, people have made up their mind to opt for the vegan options for health, environmental, or ethical reasons. When done in the perfect way, a vegan diet can help in showcasing a wide array of health benefits, for example, better control over blood sugar and a slimmer waistline. However, when a diet is based entirely on plant derivatives, it can result in a nutrient deficiency in various cases.

Veganism is being defined as a simple way of living that aims at excluding all major forms of animal cruelty and exploitation, whether for daily food, clothing, or some other purpose. For all these reasons, this diet does not include any form of animal products, such as eggs, dairy, and meat. It has been found that all those people who tend to practice veganism are thinner and also comes with a lower BMI or body mass index when compared with non-vegans. This can easily explain the primary reason why the majority of the people are turning to this form of diet as the only way for losing extra weight.

Adopting a vegan diet can help keep the blood sugar level under proper check and type 2 diabetes. According to some studies, vegans tend to benefit from the lower levels of blood sugar, higher sensitivity to insulin, and about 77% lower risk of developing diabetes than the non-vegans. The majority of the advantages can be easily explained by the increased consumption of fiber, which can blunt the blood sugar response. Several observational studies reported that vegans could have a 74% lower risk of having increased blood pressure along with a 43% lower risk of suffering from any chronic heart disease.

There is a specific food group that you will need to omit for following a vegan diet. This group of foods includes:

- **Poultry and meat:** Lamb, beef, veal, pork, organ meat, chicken, wild meat, goose, turkey, duck, etc.
- **Dairy:** Yogurt, butter, milk, cheese, ice cream, cream, etc.

- **Seafood and fish:** Fish of all types, squid, anchovies, calamari, shrimp, lobster, mussels, crab, etc.

- **Eggs:** From ostrich, chicken, fish, quail, etc.

- **Products from bees:** Bee pollen, honey, royal jelly, etc.

- **Ingredients based on animals:** Casein, whey, egg white albumen, shellac, lactose, gelatin, L-cysteine, isinglass, omega-3 fatty acids derived from fish, vitamin D3 derived from animals, and carmine.

You can opt for alternatives such as legumes, seitan, tofu, seeds, nut butter, nuts, veggies, fruits, whole grains, etc. The cooking style will remain the same, and the only difference will be the ingredients you are going to use.

Chapter 1: Breakfast Recipes

Breakfast is an essential meal of the day that needs to be fulfilled properly. Here are some vegan breakfast recipes for you.

Shamrock Sandwich

Total Prep & Cooking Time: Fifteen minutes

Yields: One serving

Nutrition Facts: Calories: 562 | Protein: 23g | Carbs: 43.2g | Fat: 31g | Fiber: 8.3g

Ingredients:

- One sausage patty (vegan)
- One cup of kale
- Two tsps. of olive oil (extra virgin)
- One tbsp. of pepitas
- Pepper and salt (according to taste)

For the sauce:

- One tbsp. of vegan mayonnaise
- One tsp. of jalapeno (chopped)
- One-fourth tsp. of paprika (smoky)

Other ingredients:

- One-fourth of an avocado (sliced)
- One toasted English muffin

Method:

1. Start by toasting the English muffin and keep it aside.
2. Take a sauté pan and drizzle some oil in it. Add the sausage patty and cook for two minutes on each side.
3. Add pepitas and kale to the hot pan. Add pepper and salt for adjusting the taste. When the kale gets soft, and the patty gets browned, remove the pan from heat.
4. Combine the spicy sauce.
5. Assemble the sandwich with sauce on the muffins and add the patty, avocado, pepitas, and kale.

Breakfast Burrito

Total Prep & Cooking Time: Thirty minutes

Yields: Two servings

Nutrition Facts: Calories: 618 | Protein: 20.3g | Carbs: 113g | Fat: 13.7g | Fiber: 21g

Ingredients:

- Three-fourth cup of rice (rinsed)
- Two cups of water
- One-fourth tsp. of salt
- One tbsp. of lime juice
- Half cup of cilantro (chopped)

For onions and hash browns:

- Half red onion
- Four red potatoes
- Two tbsp. of olive oil
- One-fourth tsp. of each
    - Black pepper (ground)
    - Salt

For black beans:

- One cup of black beans (cooked)
- One-fourth tsp. of each
    - Chili powder
    - Cumin powder
    - Garlic powder

For the avocado slaw:

- One avocado
- Two tbsps. of lime juice
- One cup of green cabbage (sliced thinly)

- One jalapeno (sliced)
- Half tsp. of each
    - Black pepper
    - Salt

For serving:

- Two flour tortillas
- Half avocado (ripe, sliced)
- One-fourth cup of salsa

Method:

1. Boil water, salt, and rice in a pan. Simmer the mixture for twenty minutes until the rice turns fluffy. Drain and keep aside.

2. Heat oil in a pan. Chop the potatoes into small pieces and slice the onions in rings. Add the potatoes along with the onions to the pan. Add pepper and salt for seasoning; toss the mixture for five minutes. Keep aside.

3. Prepare the beans in a saucepan over a medium flame.

4. For making the slaw, mix lime juice and avocado in a bowl. Mash the avocado and mix. Add jalapeno, cabbage, and toss for combining. Add pepper and salt for seasoning.

5. Add cilantro and lime juice to the rice and combine using a fork.

6. Warm the tortillas in a pan or microwave for twenty seconds.

7. Add the prepared fillings to the tortillas in order of your choice and top with salsa; add sliced avocado from the top. Roll the tortillas and slice in half.

8. Serve with extra black beans and potatoes by the side.

Gingerbread Waffles

Total Prep & Cooking Time: Fifteen minutes

Yields: Six servings

Nutrition Facts: Calories: 170 | Protein: 4.2g | Carbs: 27.3g | Fat: 4.7g | Fiber: 3.9g

Ingredients:

- One cup of flour
- One tbsp. of flax seeds (ground)

- Two tsps. of baking powder
- One-fourth tsp. of each
    - Salt
    - Baking soda
- One tsp. of cinnamon (ground)
- One and a half tsps. of ginger (ground)
- Four tbsps. of brown sugar
- One cup of any non-dairy milk
- Two tbsps. of apple cider vinegar
- Two tbsps. of molasses
- One and a half tbsps. of olive oil

Method:

1. Preheat a waffle iron and grease it.
2. Put the dry ingredients in a mixing bowl and combine well.
3. Combine the wet ingredients in a medium mixing bowl or jug. Mix well until properly combined.
4. Add the mixture of wet ingredients into the dry mixture and mix well. The batter needs to be thick. In case the batter is excessively thick, you can add two tbsps. of non-dairy milk to the batter and mix.
5. Pour the batter in batches in the waffle iron and cook until steam stops to come out from the sides.
6. Open the waffle iron and take out the waffle carefully.
7. Serve warm.

Green Chickpeas And Toast

Total Prep & Cooking Time: Thirty minutes

Yields: Two servings

Nutrition Facts: Calories: 189 | Protein: 11.3g | Carbs: 27.1g | Fat: 3.2g | Fiber: 9.3g

Ingredients:

- Two tbsps. of olive oil
- Three shallots (diced)
- One-fourth tsp. of paprika (smoked)
- Two cloves of garlic (diced)
- Half tsp. of each
  - Sweet paprika
  - Salt
  - Cinnamon
  - Sugar
- Black pepper (according to taste)
- Two tomatoes (skinned)
- Two cups of chickpeas (cooked)
- Four crusty bread slices

Method:

1. Take a medium pan and heat oil in it.
2. Add the diced shallots to the oil and stir-fry. Add garlic to the pan. Cook for five minutes until shallots turn translucent.
3. Add spices to the pan and combine well with garlic and shallots. Stir for two minutes.
4. Add the tomatoes to the pan and squash them using a spoon or spatula. Add four tbsps. of water to the pan and simmer for twelve minutes.
5. Add cooked chickpeas and mix well. Add pepper, sugar, and salt.
6. Serve the cooked chickpeas on bread slices.

Asparagus And Tomato Quiche

Total Prep & Cooking Time: One hour and twenty minutes

Yields: Eight servings

Nutrition Facts: Calories: 219 | Protein: 4.1g | Carbs: 20.6g | Fat: 11.7g | Fiber: 3g

Ingredients:

- Two cups of flour
- Half tsp. of salt
- Half cup of non-dairy butter
- Two tbsps. of water (ice cold)

For filling:

- One tbsp. of coconut oil
- One cup of asparagus (chopped)
- One-fourth cup of onion (minced)
- Three tbsps. of each
    - Sun-dried tomatoes (chopped)
    - Nutritional yeast
    - Basil (chopped)
- One block of tofu (firm)
- One tbsp. of each
    - Flour
    - Non-dairy milk
- One tsp. of each
    - Minced onion (dehydrated)
    - Mustard
- Two tsps. of lemon juice
- Half tsp. of each

- Salt
- Turmeric
- Liquid smoke

Method:

1. Spray a pie pan with oil and keep aside. Preheat your oven at 180 degrees Celsius.
2. Mix salt along with flour in a bowl. Add non-dairy butter along with cold water to the flour. Knead the dough on a working surface.
3. Press the dough on the pan. Bake the dough in the preheated oven for ten minutes.
4. Heat some oil in a pan and start adding asparagus, tomato, and onion—Cook for three minutes.
5. Combine onion, tofu, yeast, flour, non-dairy milk, lemon juice, liquid smoke, and salt in a blender.
6. Combine the mixture of asparagus with the tofu mixture.
7. Add the filling on the baked crust and smoothen the top.
8. Bake for half an hour.
9. Serve warm.

Breakfast Bowl

Total Prep & Cooking Time: One hour and twenty-five minutes

Yields: Two servings

Nutrition Facts: Calories: 350 | Protein: 7.2g | Carbs: 54g | Fat: 11.3g | Fiber: 9.3g

Ingredients:

- Two small sweet potatoes
- Cinnamon (ground, according to taste)
- Two tbsps. of each
  - Chopped nuts
  - Raisins
  - Almond butter

Method:

1. Preheat your oven at 180/160 degrees Celsius. Wash the potatoes and dry them using a kitchen towel. Use a fork for poking holes in the potatoes and wrap them using aluminum foil. Bake the potatoes for eighty minutes. Allow the potatoes to cool down before peeling.

2. Peel the baked potatoes and mash them with cinnamon.

3. Top with chopped nuts and raisins. Drizzle some almond butter from the top and serve.

Tofu Pancakes

Total Prep & Cooking Time: Twenty minutes

Yields: Six servings

Nutrition Facts: Calories: 370 | Protein: 11.2g | Carbs: 46.3g | Fat: 13.2g | Fiber: 7.9g

Ingredients:

- Fifty grams of Brazil nuts
- Three bananas (sliced)
- Three-hundred grams of raspberries

- Maple syrup (for serving)

For batter:

- Four-hundred grams of firm tofu
- Two tsps. of each
    - Lemon juice
    - Vanilla extract
- Four-hundred ml of almond milk
- One tbsp. of vegetable oil
- Two cups of buckwheat flour
- Four tbsps. of sugar
- Two tsps. of mixed spice (ground)
- One tbsp. of baking powder

Method:

1. Preheat your oven at 160 degrees Celsius. Cook the nuts by scattering them in a tray for five minutes. Chop the nuts.

2. Add vanilla, tofu, almond milk, and lemon juice in a deep bowl; blend the mixture using a stick blender. Add oil to the mixture and blend again.

3. Take a large bowl and combine the dry ingredients; add one tsp. of salt and combine. Add the mixture of tofu and combine.

4. Heat a pan and add one tsp. oil in it. Make sure that the pan is not excessively hot.

5. Use a large spoon for dropping three spoons of batter in the pan. Swirl the pan for making the pancake even—Cook for two minutes on each side. Repeat the same for the remaining batter.

6. Serve with berries, bananas, nuts, and drizzle some maple syrup from the top.

Chapter 2: Side Dish Recipes

Side dish plays a profound role in any proper meal. I have included some tasty and easy vegan side dish recipes in this section.

Baked Beans

Total Prep & Cooking Time: Five hours and twenty minutes

Yields: Ten servings

Nutrition Facts: Calories: 249 | Protein: 11.2g | Carbs: 45.3g | Fat: 2.9g | Fiber: 13.7g

Ingredients:

- Sixteen ounces of navy beans (dry)
- Six cups of water
- Two tbsps. of olive oil
- Two cups of sweet onion (chopped)
- One garlic clove (minced)
- Four cans of tomato sauce
- One-fourth cup of brown sugar
- Half cup of molasses
- Three tbsps. of cider vinegar
- Three bay leaves
- One tsp. of mustard (dry)
- One-fourth tsp. of each
    - Black pepper (ground)
    - Nutmeg (ground)
    - Cinnamon (ground)

Method:

1. Add water and beans in a pot and boil the mixture. Lower the flame and cook for one hour. Cook until the beans are tender. Drain the beans and keep aside.

2. Preheat your oven to 160 degrees Celsius.

3. Take an iron skillet and heat oil in it. Add onions in the oil. Cook for two minutes. Add garlic to the pan.

4. Combine the onion mixture with the cooked beans. Add tomato sauce, molasses, vinegar, brown sugar, pepper, bay leaves, cinnamon, mustard, and nutmeg. Mix well.

5. Cover the dish and bake for three hours. Stir in between.

6. Remove the cover and bake for forty minutes.

Baked Potato Wedges

Total Prep & Cooking Time: Fifty-five minutes

Yields: Four servings

Nutrition Facts: Calories: 234 | Protein: 5.1g | Carbs: 42.6g | Fat: 4.3g | Fiber: 8.9g

Ingredients:

- One tbsp. of olive oil
- Eight sweet potatoes (sliced into quarters)
- Half tsp. of paprika

Method:

1. Preheat your oven at 160/180 degrees Celsius.
2. Grease a baking sheet with cooking spray.
3. Combine potatoes and paprika in a bowl. Add the potatoes to the baking sheet.
4. Bake for forty minutes.
5. Serve warm.

Black Beans And Quinoa

Total Prep & Cooking Time: Fifty minutes

Yields: Ten servings

Nutrition Facts: Calories: 143 | Protein: 8.7g | Carbs: 25.6g | Fat: 1.2g | Fiber: 8.7g

Ingredients:

- One tsp. of vegetable oil
- One large onion (chopped)
- Three garlic cloves (chopped)
- Three-fourth cup of quinoa
- Two cups of vegetable stock
- One tsp. of cumin (ground)
- One-fourth tsp. of cayenne powder
- One cup of corn kernels (frozen)
- Two cans of black beans (rinsed)
- Half cup of cilantro (chopped)
- Pepper and salt (according to taste)

Method:

1. Take a medium pan and heat oil in it. Add garlic and onion to the pan. Cook for ten minutes until browned.

2. Add quinoa to the pan and mix well. Cover the mixture with vegetable stock. Season with salt, pepper, and cayenne. Boil the mixture. Cover the pan and simmer for twenty minutes until quinoa gets tender.

3. Add the corn kernels to the pan and simmer for five minutes.

4. Add cilantro and black beans to the mixture.

5. Serve hot.

Roasted Lemon Garlic Broccoli

Total Prep & Cooking Time: Twenty-five minutes

Yields: Six servings

Nutrition Facts: Calories: 48.3 | Protein: 3g | Carbs: 6.9g | Fat: 1.8g | Fiber: 3g

Ingredients:

- Two heads of broccoli (separate the florets)
- Two tsps. of olive oil (extra virgin)
- One tsp. of salt
- Half tsp. of black pepper (ground)
- One garlic clove (minced)
- One-fourth tsp. of lemon juice

Method:

1. Preheat your oven at 180/200 degrees Celsius.
2. Toss the florets of broccoli with olive oil in a bowl. Add pepper, garlic, and salt. Spread the coated broccoli florets on a baking sheet.
3. Bake for twenty minutes.
4. Add lemon juice from the top and serve warm.

Spicy Tofu

Total Prep & Cooking Time: Twenty minutes

Yields: Four servings

Nutrition Facts: Calories: 305.2 | Protein: 20.1g | Carbs: 15.4g | Fat: 19.3g | Fiber: 5.2g

Ingredients:

- Three tbsps. of peanut oil
- One red onion (sliced)
- One pound of tofu (firm, cubed)
- One bell pepper (sliced)
- One chili pepper (chopped)
- Three garlic cloves (crushed)
- One-third cup of hot water
- Two tbsps. of each
    - Soy sauce
    - White vinegar
- One tsp. of cornstarch
- One tbsp. of each
    - Red pepper flakes (crushed)
    - Brown sugar

Method:

1. Take a wok and heat oil in it. Add the tofu to the oil and keep cooking until browned. Add bell pepper, onion, garlic, and chili pepper. Mix well and cook for five minutes.

2. Whisk vinegar, soy sauce, red pepper flakes, brown sugar, and cornstarch in a bowl.

3. Add the mixture of vinegar to the wok and toss well for coating. Simmer the mixture for five minutes.

4. Serve hot.

Spanish Rice

Total Prep & Cooking Time: Forty minutes

Yields: Four servings

Nutrition Facts: Calories: 267.2 | Protein: 4.7g | Carbs: 42.7g | Fat: 5.6g | Fiber: 3g

Ingredients:

- Two tbsps. of vegetable oil
- One cup of white rice (uncooked)
- One onion (chopped)
- Half bell pepper (chopped)
- Two cups of water
- One can of green chilies and diced tomatoes
- Two tsps. of chili powder
- One tsp. of salt

Method:

1. Take a deep skillet and heat oil in it. Add onion, rice, and bell pepper to the skillet. Sauté until onions are soft and rice gets browned.
2. Add tomatoes and water to the skillet. Add salt and chili powder.
3. Simmer the mixture for thirty minutes and cover the skillet.
4. Serve hot.

Pepper And Lemon Pasta

Total Prep & Cooking Time: Twenty minutes

Yields: Eight servings

Nutrition Facts: Calories: 232.8 | Protein: 8.5g | Carbs: 41g | Fat: 4.6g | Fiber: 3.6g

Ingredients:

- One pound of spaghetti
- Two tbsps. of olive oil
- One tbsp. of basil (dried) Three tbsps. of lemon juice
- Black pepper (ground, according to taste)

Method:

1. Take a large pot and boil water in it with light salt. Add the pasta to the pot and cook for ten minutes. Drain the pasta.
2. Combine lemon juice, black pepper, lemon juice, and basil in a bowl.
3. Add the lemon mixture to the cooked pasta and toss it properly.
4. Serve cold or hot.

Chapter 3: High Protein Recipes

The vegan diet is rich in proteins, as it is mainly composed of plant compounds. Here are some easy to make vegan high protein recipes for you.

Kale Salad With Spicy Tempeh Bits And Chickpeas

Total Prep & Cooking Time: Forty-five minutes

Yields: Four servings

Nutrition Facts: Calories: 473 | Protein: 25g | Carbs: 41g | Fat: 27.1g | Fiber: 17.3g

Ingredients:

- Eight ounces of tempeh
- One-fourth cup of vegetable oil
- One-fourth tsp. of salt
- Two tsps. of each
    - Garlic powder
    - Onion powder
    - Sweet paprika
- One tsp. of each
    - Lemon pepper
    - Chili powder
- One-eighth tsp. of cayenne powder

For salad:

- 400 grams of kale (chopped)
- One cup of carrots (shredded)
- One can of chickpeas
- Two tbsps. of sesame seeds

For dressing:

- Half cup of rice vinegar

- One-fourth cup of soy sauce
- Two tbsps. of sesame oil
- One tbsp. of ginger (grated)

Method:

1. Boil water with salt in a large pot. Blanch the kale for thirty seconds. Wash blanched kale under running water. Squeeze out excess water from kale and keep aside.
2. Preheat your oven to 180 degrees Celsius.
3. Mix all the spices for tempeh in a mixing bowl.
4. Cut tempeh into very thin slices.
5. Dip each tempeh slice in oil and arrange them on a baking sheet. Line the tray using parchment paper. Sprinkle the mix of spices from the top. Coat well.
6. Bake tempeh for twenty minutes until crispy and brown in color.
7. Take a large bowl and combine the listed ingredients for the salad.
8. Combine the ingredients of dressing in a jar and shake well.
9. Pour the dressing over the prepared salad. Toss well. Make sure the salad gets coated properly with the dressing.
10. Crumble the slices of tempeh over the salad.

Protein Bars

Total Prep & Cooking Time: One hour and twenty minutes

Yields: Ten servings

Nutrition Facts: Calories: 292 | Protein: 13g | Carbs: 37.9g | Fat: 9.6g | Fiber: 2.3g

Ingredients:

For crust:

- Two cups of oat flour
- Six apricots (dried)
- One-fourth cup of each
    - Brown rice syrup
    - Cocoa powder

For layer:

- One cup of oat flour
- Half cup of each
    - Vegan protein powder (chocolate)
    - Rolled oats

- One-fourth tsp. of salt
- Two tbsps. of each
    - Hemp seeds (hulled)
    - Chia seeds
- Half cup of almond butter
- One-fourth cup of any vegan sweetener
- One cup of coconut milk
- One tbsp. flax seeds (ground)

Method:

1. Combine all the ingredients for the crust in a blender. Keep the dough aside.

2. Take a large bowl and mix the dry ingredients for the next layer. You can use a fork for proper mixing.

3. Mix water and flax in a bowl and keep aside until it turns into a gel.

4. Add almond butter, sweetener, coconut milk, and flax gel to the mixture of dry ingredients. Use a fork for mixing properly.

5. Pour the mixture in a food processor and blend well for making it smooth.

6. Use parchment paper for lining a baking sheet. Add the crust to the sheet and press the crust out. Add the next layer on the crust and evenly spread it out.

7. Put the baking sheet in the freezer for one hour.

8. Serve by cutting bars of your desired size and shape.

Tofu And Spinach Scramble

Total Prep & Cooking Time: Thirty minutes

Yields: Two servings

Nutrition Facts: Calories: 319 | Protein: 22.1g | Carbs: 11.4g | Fat: 22g | Fiber: 6.5g

Ingredients:

- Fourteen ounces of tofu (firm, cut into cubes of half-inch)
- Half tsp. of turmeric (ground)
- Black pepper and kosher salt (to taste)
- One-eighth tsp. of cayenne powder (ground)
- Two tbsps. of olive oil (extra virgin)
- Three scallions (sliced)
- Five ounces of spinach (chopped)
- Two tsps. of lemon juice
- One cup of grape tomatoes (halved)
- Half cup of basil (chopped)

Method:

1. Mix turmeric, tofu, one-fourth tsp. of salt, cayenne, and half tsp. of black pepper in a bowl. Toss the ingredients for mixing properly.

2. Take a large skillet and heat oil in it; add the scallions and stir for about one minute. Add tofu mixture and cook for five minutes until the tofu gets browned.

3. Add lemon juice, spinach, and half tsp. of salt to the tofu. Cook for one minute until the spinach wilts. Add tomatoes and stir for one minute.

4. Remove the skillet from heat and add basil.

5. Serve hot.

Vegan Tacos

Total Prep & Cooking Time: Twenty minutes

Yields: Six servings

Nutrition Facts: Calories: 402 | Protein: 29g | Carbs: 71.2g | Fat: 5.4g | Fiber: 20.3g

Ingredients:

- One tsp. of vegetable oil
- Half onion (diced)
- Two tsps. of jalapeno (chopped)
- Twelve ounces of soy chorizo (remove the casing)
- Sixteen ounces of refried black beans
- Twelve tortillas (corn)
- Cilantro (chopped)

Method:

1. Take a skillet and heat oil in it. Add onion and jalapeno to the skillet—Cook for ten minutes. Add chorizo and cook for five minutes.

2. Take a small pan and cook the beans on low heat.

3. Arrange two tortillas for making six tacos in total.

4. Spread the beans on the tortillas; top the beans with the mixture of chorizo. Serve with cilantro from the top.

Grilled Tofu Steaks And Spinach Salad

Total Prep & Cooking Time: One hour

Yields: Two servings

Nutrition Facts: Calories: 154 | Protein: 22g | Carbs: 8g | Fat: 11.3g | Fiber: 9.3g

Ingredients:

For tofu steak:

- Half block of tofu (firm)

- One tbsp. of soy sauce
- One tsp. of each
    - Miso paste
    - Tomato paste
    - Olive oil
- Half tsp. of maple syrup
- One-fourth cup of breadcrumbs

For spinach salad:

- Two cups of baby spinach
- One tbsp. of each
    - Olive oil (extra virgin)
    - Pine nuts
    - Lemon juice
- Pinch of salt
- Pinch of black pepper (ground)

Method:

1. Cut the tofu block in half and squeeze out any excess water. Make sure you do not break the block of tofu. Use paper towels for drying the tofu.

2. Cut the tofu in size and shape of your choice.

3. Take a small mixing bowl and combine tomato paste, soy sauce, olive oil, sesame, miso paste, and maple syrup. Mix until the sauce is smooth.

4. Spread breadcrumbs in a shallow dish.

5. Dip the pieces of tofu in the prepared sauce and then coat them in breadcrumbs. Repeat for the remaining tofu.

6. Grease a grill pan with some olive oil. Add the tofu steaks and cook for fifteen minutes on each side. Cook until both sides are browned.

7. For the salad, mix the listed ingredients in a medium-sized mixing bowl. Toss the ingredients properly.

8. Serve the tofu steaks with spinach salad by the side.

Corn, Quinoa, And Edamame Salad

Total Prep & Cooking Time: Two hours and ten minutes

Yields: Four servings

Nutrition Facts: Calories: 130 | Protein: 18g | Carbs: 13g | Fat: 5.8g | Fiber: 3.1g

Ingredients:

- One cup of corn kernels (frozen)
- Two cups of shelled edamame
- Half cup of cooked quinoa
- One green onion (sliced)
- Half sweet bell pepper (diced)
- Two tbsps. of cilantro (chopped)
- One and a half tbsps. of olive oil
- One tbsp. of each
    - Lime juice
    - Lemon juice
- One-fourth tsp. of each
    - Salt
    - Thyme (dried)
    - Chili powder
    - Black pepper (ground)

Method:

1. Boil corn and edamame in water with a little bit of salt. Drain and keep aside.

2. Take a bowl and mix corn, edamame, quinoa, bell pepper, green onion, and cilantro.

3. Whisk together lemon juice, olive oil, lime juice, chili powder, salt, thyme, and black pepper in a small bowl.

4. Pour the dressing all over the salad. Mix well.

5. Chill in the refrigerator for two hours.

Lentil Soup

Total Prep & Cooking Time: Fifty-five minutes

Yields: Six servings

Nutrition Facts: Calories: 230 | Protein: 9.2g | Carbs: 31.2g | Fat: 8.6g | Fiber: 11.3g

Ingredients:

- Two tbsps. of olive oil (extra virgin)
- One onion (diced)
- Two carrots (diced)

- Two stalks of celery (diced)
- One bell pepper (diced)
- Three garlic cloves (minced)
- One tbsp. of cumin
- One-fourth tsp. of paprika
- One tsp. of oregano
- Two cups of tomatoes (diced)
- Two cans of green lentils (rinsed)
- Eight cups of vegetable stock
- Half tsp. of salt
- Cilantro (for garnishing)
- One ripe avocado (diced)

Method:

1. Heat some oil in a pot. Start adding bell pepper, onion, carrots, and celery to the pot. Sauté the veggies for five minutes until tender; add paprika, cumin, garlic, and oregano. Mix well.

2. Add chilies, tomatoes, stock, salt, and lentils; simmer the mixture for forty minutes. Add pepper and salt according to taste.

3. Serve with avocado and cilantro from the top.

Chapter 4: Dessert Recipes

Everyone loves to have some dessert after having their meals. Desserts can be vegan too. So in this section, I have included some tasty dessert recipes that you can make easily.

Chocolate Pudding

Total Prep & Cooking Time: Forty-five minutes

Yields: Two servings

Nutrition Facts: Calories: 265.1 | Protein: 8.3g | Carbs: 52.2g | Fat: 4.6g | Fiber: 4.9g

Ingredients:

- Three tbsps. of cornstarch
- Two tbsps. of water
- Two cups of soy milk
- One-fourth tsp. of vanilla extract
- One-fourth cup of white sugar
- One cup of cocoa powder

Method:

1. Take a small bowl and mix water and cornstarch for forming a fine paste.
2. Take a saucepan and heat it over medium flame. Add soy milk, sugar, vanilla, a mixture of cornstarch, and cocoa. Stir the mixture until it starts boiling. Keep cooking until the mixture gets thick.
3. Allow the pudding to cool for five minutes.
4. Chill in the fridge for twenty minutes.

Orange Cake

Total Prep & Cooking Time: Forty-five minutes

Yields: Sixteen servings

Nutrition Facts: Calories: 147.3 | Protein: 1.9g | Carbs: 21.6g | Fat: 6.3g | Fiber: 0.8g

Ingredients:

- One large-sized onion (peeled)
- Two cups of flour
- One cup of white sugar
- Half cup of vegetable oil
- One and a half tsps. of baking soda
- Half tsp. of salt

Method:

1. Preheat your oven to 190 degrees Celsius. Grease a baking pan with some oil.

2. Blend the orange in a food processor until it gets completely liquefied.

3. Combine orange juice, sugar, flour, vegetable oil, baking soda, and salt. Pour the cake batter into prepared baking pan.

4. Bake the cake for thirty minutes.

Notes:

- In case you do not want to make orange juice at home, you can use orange juice from the store.

- This cake can be converted into a plain cake by omitting orange juice and using soy milk along with rice milk.

Pumpkin Tofu Pie

Total Prep & Cooking Time: Two hours

Yields: Eight servings

Nutrition Facts: Calories: 229.3 | Protein: 4.7g | Carbs: 33.6g | Fat: 8.6g | Fiber: 3.7g

Ingredients

- Ten ounces of silken tofu (drained)
- One can of pumpkin puree
- Three-fourth cup of white sugar
- Half tsp. of salt
- One tsp. of cinnamon (ground)
- One-fourth tsp. of ginger (ground)
- One-eighth tsp. of cloves (ground)
- One pie crust (unbaked)

Method:

1. Preheat the oven at 220/230 degrees Celsius.
2. Add pumpkin puree, tofu, cinnamon, salt, sugar, clove, and ginger in a food processor. Blend the ingredients until smooth.
3. Pour over the blended mixture into the crust.
4. Bake the pie for fifteen minutes and then reduce the temperature to 175 degrees Celsius. Bake again for forty minutes.
5. Let the pie cool down.
6. Serve at room temperature.

Note: If you are allergic to certain ingredients, check the ingredients of the pie crust.

Vegan Brownie

Total Prep & Cooking Time: Fifty minutes

Yields: Sixteen servings

Nutrition Facts: Calories: 254.3 | Protein: 2.6g | Carbs: 38.3g | Fat: 13.6g | Fiber: 2.8g

Ingredients:

- Two cups of flour
- Two cups of white sugar
- Three-fourth cup of cocoa powder (unsweetened)
- One tsp. of baking powder
- Half tsp. of salt
- One cup of each
    - Vegetable oil
    - Water
- One and a half tsps. of vanilla extract

Method:

1. Preheat your oven at 160/175 degrees Celsius.

2. Take a large bowl and combine sugar, flour, cocoa powder, salt, and baking powder. Add vegetable oil, vanilla extract, and water to the mixture. Mix well for making a smooth batter.

3. Spread the brownie mix in a baking pan.

4. Bake the brownie for thirty minutes.

5. Allow the brownie to cool for ten minutes.

6. Cut in squares and serve.

Vegan Cupcake

Total Prep & Cooking Time: Twenty-five minutes

Yields: Eighteen servings

Nutrition Facts: Calories: 150.2 | Protein: 1.9g | Carbs: 21.6g | Fat: 6.3g | Fiber: 0.9g

Ingredients:

- One tbsp. of cider vinegar
- Two cups of almond milk
- Two and a half cups of flour
- One cup of white sugar
- Two tsps. of baking powder
- Half tsp. of each
    - Salt
    - Baking soda
    - Coconut oil (warmed)
- One and a half tsps. of vanilla extract

Method:

1. Preheat the oven at 160/175 degrees Celsius. Grease eighteen muffin cups using some oil.
2. Mix cider vinegar and almond milk in a bowl. Allow it to stand for five minutes until the mixture gets curdled.
3. Take a bowl and combine sugar, salt, baking powder, flour, and baking soda.
4. Take another bowl and combine coconut oil, vanilla, and almond milk mixture. Add this mixture to the mixture of dry ingredients.
5. Divide the batter into muffin cups—Bake for twenty minutes.
6. Allow the cupcakes to sit for ten minutes.
7. Serve with the desired frosting from the top.

Vanilla Cake

Total Prep & Cooking Time: Fifty minutes

Yields: Eight servings

Nutrition Facts: Calories: 277.9 | Protein: 3.5g | Carbs: 44.2g | Fat: 9.2g | Fiber: 0.9g

Ingredients:

- One cup of soy milk
- One tbsp. of apple cider vinegar
- One and a half cup of flour
- One cup of white sugar
- One tsp. of each
    - Baking powder
    - Baking soda
- Half tsp. of salt
- One-third cup of canola oil
- One-fourth tsp. of almond extract
- One tbsp. of vanilla extract
- One-fourth cup of water

Method:

1. Preheat your oven at about 160/175 degrees Celsius. Grease a baking pan with some oil.

2. Mix vinegar and soy milk in a large cup.

3. Combine sugar, flour, salt, baking soda, and baking powder in a bowl.

4. Add lemon juice, canola oil, vanilla extract, almond extract, and water to the mixture of soy milk. Stir the mixture of soy milk into the mixture of flour. Mix well until there is no lump.

5. Pour the cake batter in the baking dish.

6. Bake in the preheated oven for thirty-five minutes.

Miracle Fudge

Total Prep & Cooking Time: One hour and ten minutes

Yields: Twenty-four servings

Nutrition Facts: Calories: 77.3 | Protein: 0.9g | Carbs: 4.9g | Fat: 6.2g | Fiber: 1.2g

Ingredients:

- Half cup of cocoa (unsweetened)
- One cup of maple syrup
- One tsp. of vanilla extract
- One pinch of salt
- One-third cup of each
    - Chopped walnuts
    - Coconut oil (melted)
- One tsp. of cocoa powder (unsweetened, for dusting)

Method:

1. Add half cup of cocoa powder in a bowl along with maple syrup; give it a stir. Add vanilla extract and salt. Add melted coconut oil and combine well.

2. Add walnuts in a pan and toast them for one minute.

3. Add the toasted walnuts to the fudge. Mix well.

4. Pour the mixture of fudge into a silicone mold. Smoothen the top.

5. Wrap the silicone mold using plastic wrap and put it in the freezer for thirty minutes. Take out the fudge pieces from the mold and dust with cocoa powder from the top.

6. Serve cold.

Chapter 5: Sauces & Dips Recipes

Sauces and dips are important components of any meal that can make the food tastier. Here are some vegan sauces and dips recipes for you.

Tomato Jam

Total Prep & Cooking Time: Forty-five minutes

Yields: One full cup

Nutrition Facts: Calories: 34 | Protein: 0.2g | Carbs: 8.6g | Fat: 0.1g | Fiber: 0.2g

Ingredients:

- Two pounds of plum tomatoes
- One-fourth cup of coconut sugar
- Half tsp. of salt
- One-fourth tsp. of paprika
- One tsp. of vinegar (white wine)
- Black pepper (to taste)

Method:

1. Take a large pot and boil water in it; add the tomatoes to the water and boil for one minute. Remove the tomatoes and put them in an ice-water bath.

2. Peel the blanched tomatoes and chop them.

3. Add chopped tomatoes in a pot over a medium flame. Add sugar and stir the mixture—Cook for ten minutes.

4. Add pepper, salt, and paprika. Simmer for ten minutes until the jam thickens.

5. Remove from heat and add white wine vinegar. Serve with crackers, burgers, toasts, etc.

Walnut Kale Pesto

Total Prep & Cooking Time: Thirty minutes

Yields: One small bowl

Nutrition Facts: Calories: 240 | Protein: 3.6g | Carbs: 2.6g | Fat: 22.3g | Fiber: 0.8g

Ingredients:

- Half bunch of kale (chopped)
- Half cup of walnuts (chopped)
- Two garlic cloves (minced)
- One-fourth cup of yeast
- One cup of olive oil
- Three tbsps. of lemon juice
- Pepper and salt (for seasoning)

Method:

1. Take a pot and boil water in it. Add kale to the pot with one tsp. of salt—Cook for five minutes.
2. Add kale, garlic, walnuts, olive oil, yeast, along with lemon juice in a food processor. Add pepper and salt according to your taste. Blend well.

Ranch Dressing

Total Prep & Cooking Time: Thirty minutes

Yields: One small bowl

Nutrition Facts: Calories: 92 | Protein: 0.1g | Carbs: 1g | Fat: 9.1g | Fiber: 0.4g

Ingredients:

- One cup of vegan mayo
- Half tsp. of each
  - Onion powder
  - Garlic powder
- One-fourth tsp. of black pepper (ground)
- Two tsps. of parsley (chopped)
- One tbsp. of dill (chopped)
- Half cup of soy milk (unsweetened)

Method:

1. Mix the listed ingredients in a medium mixing bowl.
2. If you want the dressing to be thin, add a bit of almond milk. Allow the dressing to sit for a few minutes.
3. Chill the prepared dressing.

Notes:

- You can serve the dressing with any savory snacks, sandwiches, quick-bites, salads, etc.
- You can store the leftover dressing in the fridge for two days.
- You can add some ground nuts for enhancing the flavor.

PART III

Chapter 1: Identifying the Mediterranean Diet

We know that certain diets are associated with better health—this is a simple fact of life. We've seen that entire groups of people live longer based on where they live, and to some degree, a good deal of that has to come from somewhere—it has to come from something like diet or environment. In this case, the diet of the people living in the Mediterranean has been found to be incredibly healthy for people—it has been shown that people who are able to enjoy this diet, who are able to eat fresh food by the sea and enjoy the benefits that it has, are able to be far healthier than those who don't have it. That is great for them—but what is their secret?

It turns out, it's all in the lifestyle. The Mediterranean lifestyle, food, and all, is incredibly healthy for you. Studies have shown that people living in Mediterranean countries such as Greece and Italy have been found to have far less risk of death from coronary disease. Their secret is in the diet. Their diet has been shown to reduce the risk of cardiovascular disease, meaning that it is incredibly healthy, beneficial, and something that the vast majority of people in the world could definitely benefit from.

The Mediterranean diet is recommended by doctors and the World Health Organization as being not only healthy but also sustainable, meaning that it is something that is highly recommended, even by the experts. If you've found that you've struggled with weight loss, heart disease, managing your blood pressure, or anything similar to those problems, then the Mediterranean diet is for you.

When you follow this diet, you are able to bring health back to your life and enjoy the foods while doing so. It's perfect if you want to be able to enjoy your diet without having to worry about the impacts that it will have on you.

Defining the Mediterranean Lifestyle

The Mediterranean diet is quite simple. It involves eating traditional foods based on one's location. Typically, in the Mediterranean, that is a diet that is rich in veggies, fruits, whole grains, beans, and features olive oil as the fat of choice. Typically, it involves elements beyond just eating as well. While it is important to have healthy food, it is equally important to recognize that the diet encompasses the lifestyle as well. In particular, you can expect to see a few other rules come into play.

In particular, the Mediterranean diet is unique in the sense that it encourages a glass of red wine every now and then. In fact, the diet is associated with moderate drinking, enjoying red wine several times per week, always responsibly, and in contexts that will be beneficial to the drinker. If you want to be able to enjoy the Mediterranean diet and you are pregnant, or against drinking, you can do that, too—but traditionally, the red wine is included and even encouraged in moderation thanks to the antioxidants within it.

Additionally, on the Mediterranean diet, it is common to share meals with friends and family. This is essential—eating is more than just filling the body, it is nurturing the mind and relationships as well. This also comes with the added benefit of also being able to slow down eating—when you are eating the foods on this diet, you will discover that ultimately, you eat less when you're busy having a riveting conversation with someone. The fact that you are slowed down with your eating means that you will fill up sooner and realize that you didn't have to actually eat the food that you did. This means that you eat less and are, therefore managing your portions better as a result.

Finally, the Mediterranean diet focuses on physical activity. Traditionally, you would have had to go out to get the foods that you would eat each day, and that would mean that you'd need to get up, fish, garden, farm, or otherwise prepare your food. Eating locally is still a major component of this diet, as is getting up and being active. You need at least 30 minutes of activity, moderate or mild, per day. Even just walking for half an hour is better than nothing!

The Rules of the Mediterranean Diet

To eat the Mediterranean way, there are a few key factors that can guide you. If you know what you are doing, you can eat well without having to sacrifice flavor for health, and that matters immensely. When you look at the Mediterranean diet closely, you see that there are several tips that will help you to recognize what you need to do to stick to your diet.

Eating fruits and veggies

First, make sure that the bulk of your calories come from fruits and vegetables. You should be eating between 7 and 10 servings of fresh fruits and vegetables every single day—meaning that the bulk of your calories will come from there. Try to stick to locally grown foods that are fresh and in-season—they will have the highest nutritional value.

Reach for the whole grains

Yes, pasta is a major part of the diet in the Mediterranean, and you don't have to give that up entirely—but make sure that any grains that you are enjoying are whole-wheat. This allows you to enjoy foods that are high in fiber and are able to be digested differently than when you use refined carbs instead. While the refined carbs may give you instantaneous energy, they are also not nearly as good for you as whole wheat.

Using healthy fats

When it comes to flavoring or cooking your foods, you need to reach for the healthy fats first. This means choosing out foods that are cooked with olive oil instead of butter or dipping food in olive oil instead of butter. Olive oil, despite being a fat, has not been found to lead to weight gain when used in moderation. It is an incredibly healthy substitute for butter that is loaded up with all sorts of beneficial, heart-healthy antioxidants that will help your cardiovascular system.

Aim for seafood

When it comes to protein, fish, especially fresh fish, is the best choice. Fish should be consumed at least twice per week, and it should be fresh rather than frozen whenever possible. In particular, it is commonly recommended that you reach for salmon or trout, or other fatty fish because the omega-3 fatty acids within them are incredibly healthy for you, and they will serve you well. Even better, if you grill your fish, you have little cleanup.

Reduce red meat

In addition to adding more seafood to your diet, you need to cut out the red meat. The red meats in your diet are no good for you—they have been linked to inflammation that can make it harder for your cardiovascular system.

Enjoy dairy in moderation

When you are on this diet, dairy is not out of the picture entirely. While you should avoid butter, for the most part, it is a good idea for you to enjoy some low-fat Greek yogurt on occasion and add in some cheese to your diet. It is a good thing for you to enjoy these foods to ensure that you have plenty of calcium to keep your body strong.

Spices, not salt

Perhaps one of the most profound differences between most other diets and the Mediterranean diet is the lack of salt. The Mediterranean diet reaches for herbs and spices before adding in salt, meaning that you will be consuming less of it over time. Even better, you will grow to love your new foods without needing salt.

Chapter 2: Savory Mediterranean Meals

Mediterranean Feta Mac and Cheese

Ingredients

- Egg (1, beaten)
- Feta cheese (8 oz., crumbles)
- Macaroni (0.5 lb., whole-wheat)
- Olive oil (3 Tbsp.)
- Salt and pepper to taste
- Sour cream (8 oz.)

Instructions

1. Cook pasta to instructions to create al dente pasta. Drain and place pasta into baking dish. Toss in feta and oil and mix well.
2. Combine your egg and sour cream with salt and pepper. Then mix well and toss over macaroni. Combine and bake at 350F for 30 minutes.

Chickpea Stew

Ingredients

- Bay leaf (1)
- Dry chickpeas (1 c., soaked overnight and peeled)
- Garlic (1 clove, cut in half)
- Lemon to serve

- Olive oil (0.25 c.)
- Onion (1, diced)
- Salt and pepper to taste

Instructions

1. Cover chickpeas in pot with just enough water to cover them and wait to boil. Then rinse and set into clean pot. Toss in all other ingredients but the lemon with just enough water to cover nearly one inch above the beans. Simmer for 2-3 hours and serve with lemons.

Savory Mediterranean Breakfast Muffins

Ingredients

Dry ingredients

- Baking powder (1.5 tsp)
- Baking soda (o.5 tsp)
- Flour (2 c.)
- Salt (0.5 tsp)

Wet ingredients

- Egg (1 large)
- Garlic (1 clove, minced)
- Milk (1 c.)
- Sour cream (0.25 c.)
- Vegetable oil (0.25 c.)

Fillings

- Cheddar cheese (2 c., shredded)
- Feta (2.5 oz., crumbled)
- Green olives (diced, 0.5 c.)
- Green onions (0.5 c., chopped)
- Roasted red peppers (0.5 c., chopped)
- Sun dried tomatoes (diced, 0.5 c.)

Instructions

1. Combine dry ingredients in a bowl. Mix wet ingredients in separate bowl. Combine the two together and mix.
2. Toss in fillings in as few stirs as possible.
3. Place in greased or lined muffin pan, dividing to all 12 recesses.
4. Bake for 25 minutes until golden-brown and crusty at 350F.
5. Cool for 10 minutes and serve warm.

Mediterranean Breakfast Bake

Ingredients

- Artichoke hearts (14-oz. can, drained)
- Bread (6 slices whole-wheat, chopped)
- Eggs (8)
- Feta cheese (0.5 c.)
- Italian sausage (turkey or chicken—1 lb., casings removed)
- Milk (1 c.)
- Olive oil (2 Tbsp., divided)

- Onion (1, chopped)
- Spinach (5 oz.)
- Sun dried tomato (1 c., chopped)

Instructions

1. Warm 1 Tbsp. of your olive oil on moderately high heat. Cook sausage for 8 minutes until it has browned, breaking it up as it cooks. Place it in a dish when it is done.
2. Toss in additional oil, then cook onion until soft, roughly 5 minutes. Toss in spinach until wilting (1 minute).
3. Combine eggs and mix in milk, bread, tomatoes, cheese, artichokes, sausage, and finally, the spinach mix.
4. Place everything in a 2.5 quart baking dish. Let sit for an hour in fridge, or leave overnight.
5. Let casserole sit for 30 minutes after removing from fridge. Then, bake for 45 minutes at 350F until brown. Let rest 10 minutes, then serve.

Mediterranean Pastry Pinwheels

Ingredients

- Cream cheese (8-ounce package, softened)
- Pesto (0.25 c.)
- Provolone cheese (0.75 c.)
- Sun-dried tomatoes (0.5 c., chopped)
- Ripe olives (0.5 c., chopped)

Instructions

1. Unroll pastry and trim it up to create 10-inch square.
2. Mix together your cream cheese and pesto until well-combined. Then, mix in other ingredients until combined. Place mixture in even layer across pastry, up to 0.5-inch of edges. Roll and freeze for 30 minutes.
3. Cut whole roll into 16 pieces.
4. Bake at 400F until golden, roughly 15 minutes. Serve.

Chapter 3: Sweet Treats on the Mediterranean Diet

Greek Yogurt Parfait

Ingredients

- Almond butter (2 Tbsp.)
- Fresh fruit (1 Tbsp.)
- Greek Yogurt (1 c.)

Instructions

1. Mix together yogurt and 1 Tbsp. of almond butter and put in a bowl. Top with fruit.
2. Warm remaining butter in microwave for 10 minutes, then drizzle atop yogurt. Serve. You can add different toppings to change up the flavor as well.

Overnight Oats

Ingredients

- Chia seeds (1 Tbsp.)
- Greek yogurt (0.25 c.)
- Honey (1 Tbsp.)
- Milk of choice (0.5 c.)
- Old fashioned whole oats (0.5 c.)
- Vanilla extract (0.25 tsp)

Instructions

1. Mix all ingredients into a glass container and leave in fridge for at least 2 hours but preferably overnight. Serve with berries of choice or other desired toppings.

Apple Whipped Yogurt

Ingredients

- Greek yogurt (1 c.)
- Heavy cream (0.5 c.)
- Honey (1 Tbsp.)
- Unsalted butter (2 Tbsp.)
- Apples (2, cored and chopped into small bits)
- Sugar (2 Tbsp.)
- Cinnamon (1/8 tsp)
- Walnut halves (0.25 c., chopped)

Instructions

1. Using a hand mixer, mix together yogurt, honey, and honey until it creates peaks.
2. Heat up your butter in a skillet over a moderate temperature. Cook apples and 1 Tbsp. sugar in pan. Stir and cook for 6-8 minutes until soft. Then, top with the rest of sugar and cinnamon, stirring and cooking an additional 3 minutes. Take it off of the burner and let it rest for 5 minutes.
3. Serve with whipped yogurt in bowl topped with apple, then sprinkle on walnuts.

Chapter 4: Gourmet Meals on the Mediterranean Diet

Garlic-Roasted Salmon and Brussels Sprouts

Ingredients

- Brussels sprouts (6 c., trimmed and halved)
- Chardonnay (0.75 c.)
- Garlic cloves (14 large)
- Olive oil (0.25 c.)
- Oregano (2 Tbsp., fresh)
- Pepper (0.75 tsp)
- Salmon fillet (2 lbs., skin-off—cut in 6 pieces)
- Salt (1 tsp)
- Lemon wedges to serve

Instructions

1. Take two cloves of garlic and mince, combining them with oil, 1 Tbsp. of oregano, half of the salt and 1/3 of the pepper. Cut remaining cloves of garlic in halves and toss them with the sprouts. Take 3 Tbsp. of your garlic oil and toss it with the sprouts in roasting pan. Roast for 15 minutes at 450F.
2. Add wine to the remainder of the oil mixture. Then, remove it from the pan, stir veggies, and place salmon atop it all. Pour the wine mix atop it and season with remaining oregano and salt and pepper. Bake 5-10 minutes until salmon is done. Serve alongside the wedged lemon.

Walnut Crusted Salmon with Rosemary

Ingredients

- Dijon mustard (2 tsp)
- Garlic (1 clove, minced)
- Honey (0.5 tsp)
- Kosher salt (0.5 tsp)
- Lemon juice (1 tsp)
- Lemon zest (0.25 tsp.)
- Olive oil (1 tsp)
- Olive oil spray
- Panko (3 Tbsp.)
- Red pepper (0.25 tsp)
- Rosemary (1 tsp, chopped)
- Salmon (1 pound, skin removed)
- Walnuts (3 Tbsp., finely chopped)

- Parsley and lemon to garnish

Instructions

1. Mix together the mustard, lemon zest and juice, honey, salt and red pepper, and rosemary. In a separate dish, combine the panko with oil and walnuts.
2. Spread mustard across salmon and top with panko mixture. Spray fillets with cooking spray.
3. Cook until fish begins to flake at 425F, roughly 8-10 minutes. Serve with lemon and parsley.

Spaghetti and Clams

Ingredients

- Clams (6.5 lbs.)
- Olive oil (6 Tbsp.)
- White wine (0.5 c.)
- Garlic (3 cloves, sliced)
- Chiles (3, small and crumbled)
- Spaghetti (1 lb.)
- Parsley (3 Tbsp., chopped)
- Salt and pepper to personal preference

Instructions

1. Prepare clams, soaking in clean water and brushing to remove all sand.

2. Warm 2 Tbsp. of oil in large pot. Then, toss in 0.25 c. wine, 1 of the cloves of garlic, and 1 chile. Cook half of the plans at high heat with regular shaking until clams are opened. Remove opened clams and their juices to a larger bowl. Repeat process with second half of clams. Discard any that do not open.
3. Prepare pasta according to packaging to create al dente pasta. Reserve 1 c. pasta water.
4. Warm remainder of oil (2 Tbsp.) in pot over moderate heat, tossing in remainder of garlic and chile. Cook until fragrant, then place all clams and their juices into the pot, tossing to coat well. Then, toss in pasta, mixing well to combine. If necessary, add in cooking liquid. Serve and season with salt/pepper to personal preference with parsley atop.

Braised Lamb and Fennel

Ingredients

- Bay leaves (2)
- Chicken broth (3 c.)
- Cinnamon stick (1)
- Fennel (1 bulb, chopped)
- Garlic head (chopped in half)
- Lamb shoulder (3 lbs., cut into 8 pieces)
- Olive oil (2 Tbsp.)
- Onion (1, chopped
- Orange (1 with peel, cut into wedges)
- White wine (1 c.)
- Whole peeled tomatoes (14.5 oz. can)

Instructions

1. Dry lamb and season with salt and pepper to taste. Warm oil inside a Dutch oven, and sear lamb on all sides, roughly 6 minutes each side. Move lamb to plate.
2. Place fennel, garlic, and onion in the pot and cook, until browning, roughly 8 minutes. Mix in wine and boil, deglazing the pan. Reduce heat and simmer until it has reduced 50%.
3. Toss in orange, bay leaves, tomatoes, broth, and cinnamon, plus the lamb. Simmer, then cover pot and transfer to oven set to 325F. braise for 1.5-2 hours. Remove lamb and place on clean plate.
4. Strain liquid left in pot, then return it to the pot to boil until thick, roughly 30 minutes.
5. Return lamb to pot to warm. Serve.

Mediterranean Cod

Ingredients

- Black olives (0.66 c., sliced)
- Cod (4 fillets, skinless)
- Fennel seeds (1 tsp)
- Lemon (1, sliced)
- Lemon (juice of ½ lemon)
- Olive oil (6 Tbsp.)
- Onion (1, sliced)
- Parsley (1 Tbsp., chopped)
- Salt and pepper to personal preference
- Tomatoes (0.66 c., diced)

Instructions

1. Warm olive oil at a moderate temperature, sautéing the onion with a pinch of salt until translucent, roughly 10 minutes.
2. Mix in tomato and olives, tossing in the juice as well. Allow it to simmer gently for roughly 5 minutes. Toss in fennel seeds and set aside.
3. Warm the rest of the oil in another pan and fry up the cod for 10 minutes, flipping halfway through until done.
4. Toss tomato sauce over heat to warm, then mix together the parsley, and serve atop the cod with a lemon slice.

Baked Feta with Olive Tapenade

Ingredients

- Baked pita or crusty bread to serve
- Feta cheese (6 oz.)
- Garlic (2 cloves)
- Green olives (0.33 c., sliced)
- Harissa paste (3 Tbsp.)
- Olive oil (3 Tbsp.)
- Parsley (3 Tbsp., fresh chopped)
- Roasted red peppers (16-oz. jar, drained)
- Salt (0.75 tsp.)
- Tomato paste (2 Tbsp.)
- Walnuts (0.5 c., halved)

Instructions

1. In a blender, combine your peppers, 0.25 c. walnuts, harissa and tomato paste, garlic, and 0.5 tsp of your salt until mostly consistent. It doesn't have to be perfect, but should be well combined.
2. Take half of mixture into baking dish that has been sprayed with cooking spray. Top with half of your feta, then spoon the rest of the red pepper sauce atop it.
3. Top with the last of the feta and bake until bubbly, roughly 25 minutes. Broil for the last 2.
4. While that bakes, make your tapenade. This requires you to combine your remaining ingredients together.
5. Remove mixture from oven and top with tapenade. Serve immediately with crusty bread or pita chips.

Chapter 5: 30-Minutes or Less Meals

Vegetarian Toss Together Mediterranean Pasta Salad

Ingredients

- Artichoke hearts (12 oz. jar, drained)
- Balsamic vinegar (2 Tbsp.)
- Kalamata olives (12-ounce jar, drained and chopped)
- Olive oil (2 Tbsp.)
- Pasta (8 oz., wheat)
- Salt to personal preference
- Sun-dried tomatoes in oil (1.5 oz. jar, drained)

Instructions

1. Prepare pasta according to packaging.
2. Mix together olives, tomatoes, and artichoke.
3. Drain pasta and add them to a bowl with artichoke mixture. Then, top with vinegar and olive oil, mix well, and serve warm.

Vegetarian Aglio e Olio and Broccoli

Ingredients

- Olive oil (3 Tbsp.)
- Cayenne peppers (3)
- Garlic (3 cloves, sliced)
- Broccoli (1 head, prepared in florets)
- Spaghetti (7 oz. whole wheat)
- Salt to taste

Instructions

1. Boil water and prepare spaghetti according to instructions until al dente. Drain and reserve.
2. In a pan, heat up 1 Tbsp. of your olive oil at a moderate temperature, then toss in the garlic and peppers, sautéing until fragrant. Remove garlic from heat and set aside.
3. Toss broccoli into pan and cook for 4 minutes. Then toss in spaghetti, garlic, and remaining oil. Cook for an additional minute or two, then serve.

Cilantro and Garlic Baked Salmon

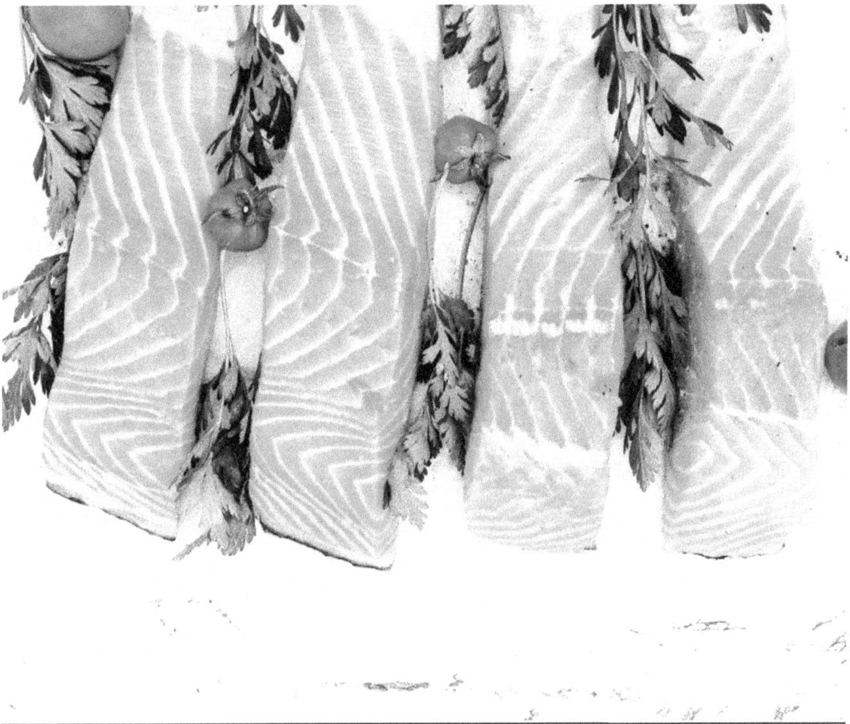

Ingredients

- Cilantro (stems trimmed)
- Garlic (4 cloves, chopped)
- Lime (0.5, cut into rounds)
- Lime juice (1 lime's worth)
- Olive oil (0.5 c.)
- Salmon fillet (2 pounds, skin removed)
- Salt to taste
- Tomato (cut into rounds)

Instructions

1. Allow salmon to come to room temp for 20 minutes while oven preheats to a temperature of 425 F.
2. While you wait, take a processor and combine garlic, cilantro, lime juice, and olive oil with a pinch of salt. Combine well.
3. Place fillet into baking pan that has been greased. Top with a light sprinkle of salt and pepper. Then spread cilantro sauce atop fillet, coating whole salmon. Top with tomato and lime.
4. Bake for 6 minutes per 0.5 inch of thickness (1-inch fillets take around 8-10 minutes). Let rest for 5-10 minutes out of the oven. Serve.

Harissa Pasta

Ingredients

- Pasta (2 cups)
- Red bell pepper (1)
- Red onion (1)
- Pine nuts (2 Tbsp.)
- Harissa paste (2 Tbsp.)

Instructions

1. Roast onions and peppers with olive oil at 400F for 20 minutes. Remove from oven and dice.
2. Prepare pasta to instructions on package. While pasta cooks, toast your pine nuts until browned in frying pan.
3. Drain pasta, leaving a touch of the water. Then, add in diced roasted veggies and harissa. Serve topped with pine nuts.

Chapter 6: 1-Hour-or-Less Meals

1 Hour Baked Cod

Ingredients

- Basil (0.5 tsp., dried)
- Bay leaf (1)
- Capers (1 small jar)
- Cod fillets (2 pounds)
- Fennel seeds (1 tsp., crushed)
- Garlic (1 clove, minced)
- Lemon juice (0.25 c., fresh)
- Olive oil (2 tsp)
- Onion (1, sliced)
- Orange juice (o.25 c., fresh)
- Orange peel (1 Tbsp.)
- Oregano (0.5 tsp., dried)
- Salt and pepper to personal preference
- White wine (1 c., dry)
- Whole tomatoes (16-oz. can, chopped and reserving juice)

Instructions

1. Warm oven to 375F.
2. In cast iron skillet, warm oil. Then, sauté your onion for 5 minutes. At this point, mix in all other ingredients but fish. Allow to simmer for 30 minutes.
3. Place fillets into skillet and top with most of the sauce. Allow to bake for 15 minutes until fish flakes.

Grilled Chicken Mediterranean Salad

Ingredients

- Artichoke hearts (0.33 c., chopped)
- Balsamic vinegar (2 Tbsp.)
- Basil (1 tsp, dried)
- Chicken breasts (3, cut into bite-sized chunks)
- Cucumber (0.75 c., diced)
- Feta cheese (0.25 c.)
- Garlic (1 clove, minced)
- Greek yogurt (2 Tbsp.)
- Green onions (0.25 c., chopped)

- Kalamata olives (3 Tbsp., sliced)
- Kosher salt (0.5 tsp)
- Lemon juice (3 Tbsp + 1 tsp.)
- Olive oil (3 Tbsp. + 2 Tbsp.)
- Onion powder (0.5 tsp)
- Parsley (0.5 tsp)
- Pesto (4 tsp)
- Pinch of red pepper
- Roasted red pepper (6 Tbsp., sliced)
- Romaine (4 c., chopped)
- Shiitake mushrooms
- Spinach (4 c., chopped)
- Tomato (0.75 c., diced)
- White wine vinegar (4 tsp)

Instructions

1. Create your salad. Each plate should have a bed of romaine and spinach, topped with cucumber, tomato, artichoke, peppers, olives, and cheese.
2. Combine your tsp of lemon juice, wine vinegar, and pesto in a jar and shake to combine. Then, add in yogurt and 2 Tbsp. oil, mixing well until well-incorporated.
3. Prepare your chicken. Let it marinade in a mixture of 3 Tbsp. lemon juice, balsamic vinegar, remaining oil, and all seasonings for at least 30 minutes. Soak some wooden skewers in water during this time.
4. Make kebabs out of chicken and mushroom, alternating bite of chicken and bite of mushroom until chicken is gone. Grill for 10 to 15 minutes until done.
5. Drizzle salad with the vinaigrette, then place a kebab atop each. Serve.

Lemon Herb Chicken and Potatoes One Pot Meal

Ingredients

- Baby potatoes (8, halved)
- Basil (3 tsp, dried)
- Bell pepper (1, seeds removed and wedged)
- Chicken thighs (4, skin and bone on)
- Garlic (4 large cloves, crushed)
- Kalamata olives (4 Tbsp., pitted)
- Lemon juice (1 lemon's worth)
- Olive oil (3 Tbsp.)
- Oregano (2 tsp, dried)

- Parsley (2 tsp, dried)
- Red onion (wedged)
- Red wine vinegar (1 Tbsp.)
- Salt (2 tsp)
- Zucchini (1 large, sliced)
- Lemons for garnish

Instructions

1. Combine juice from lemon, 2 Tbsp. olive oil, vinegar, seasonings, and garlic into dish. Pour half to reserve for later, then place chicken in half. Let sit for 15 minutes (or overnight if you would like to prep the day before)
2. Warm oven to 430F. Sear chicken in cast iron skillet in remaining olive oil, about 4 minutes per side. Drain all but 1 Tbsp. of fat.
3. Place all veggies around the thighs. Top with remaining marinade and combine well to cover everything.
4. Cover pan and bake for 35 minutes until soft and chicken is to temperature. Then, broil for 5 minutes or until golden brown. Top with olives and lemon to serve.

Vegetarian Mediterranean Quiche

Ingredients

- Butter (2 Tbsp.)
- Cheddar cheese (1 c., shredded)
- Eggs (4 large)
- Feta (0.33 c.)
- Garlic (2 cloves, minced)
- Kalamata olives (0.25 c., sliced)
- Milk (1.25 c.)
- Onion (1, diced)
- Oregano (1 tsp, dried)
- Parsley (1 tsp, dried)
- Pie crust (1, prepared)
- Red pepper (1, diced)
- Salt and pepper to personal preference
- Spinach (2 c., fresh)
- Sun dried tomatoes (0.5 c.)

Instructions

1. Soak sun-dried tomatoes in boiling water for 5 minutes before draining and chopping.
2. Prepare a pie dish with a crust, fluting the edges.
3. In a skillet, melt your butter, then cook your garlic and onions in it until they become fragrant. Combine in the red peppers for another 3 minutes until softened. Then, toss in your spinach, olives, and seasoning. Cook until the spinach wilts, about 5 minutes. Take it off of the heat and toss in your feta and tomatoes. Then, carefully place mixture into the crust, spreading it into a nice, even layer.
4. Mix milk, eggs, and half of cheddar cheese together. Pour it into the crust. Then, top with cheddar.
5. Bake for 50 minutes at 375 f. until crust is browned and egg is done.

Herbed Lamb and Veggies

Ingredients

- Bell pepper (2, any color, seeds removed and cut into bite-sized chunks)
- Lamb cutlets (8 lean)
- Mint (2 Tbsp., fresh, chopped)
- Olive oil (1 Tbsp.)

- Red onion (1, wedged)
- Sweet potato (1 large, peeled, and chunked)
- Thyme (1 Tbsp., fresh, chopped)
- Zucchini (2, chunked)

Instructions

1. Assemble your veggies onto a baking sheet and coat with oil and black pepper. Bake at 400F for 25 minutes.
2. As veggies bake, trim fat from the lamb. Then, combine the herbs with a bit of freshly ground pepper. Coat the lamb in the seasoning.
3. Remove veggies, flip, and push to one side of pan. Then, arrange your cutlets onto the baking pan as well. Bake for 10 minutes, flip, then cook an additional 10 minutes. Combine well, then serve.

Chicken and Couscous Mediterranean Wraps

Ingredients

- Parsley (1 c., fresh and chopped)
- Olive oil (3 Tbsp.)
- Garlic (2 tsp, minced)
- Salt (pinch)
- Pepper (pinch)
- Chicken tenders (1 pound)
- Tomato (1, chopped)
- Cucumber (1, chopped)
- Spinach wraps (4 1o-inch)
- Water (0.5 c)
- Mint (0.5 c., fresh chopped)
- Lemon juice (0.25 c.)
- Couscous (0.33 c.)

Instructions

1. Cook couscous in boiling water according to directions on package.
2. Mix together your lemon juice, oil, garlic, salt and pepper, mint, and parsley.
3. Coat chicken in 1 Tbsp. of your mixture from previous step and top with a pinch of salt. Cook in skillet until completely cooked, usually just a few minutes per side.
4. Wait for chicken to cool, then chop into bites.
5. Pour the remainder of your parsley mixture into the couscous with cucumbers and tomato bits.
6. Place 0.75 c. of couscous mixture into a tortilla, then spread chicken atop it, rolling them up and serving.

Sheet Pan Shrimp

Ingredients

For shrimp

- Feta cheese (0.5 c.)
- Fingerling potatoes (2 c., halved)
- Green beans (6 oz., trimmed)

- Olive oil (3 Tbsp.)
- Pepper (1 tsp)
- Red onion (1 medium, sliced)
- Red pepper (1 medium, sliced)
- Salt (1 tsp)
- Shrimp (1 lb., deveined and peeled)

For Marinade

- Garlic (1 Tbsp., minced) Oregano (0.5 tsp)
- Greek yogurt (1 c.)
- Lemon juice (2 Tbsp.)
- Paprika (0.5 tsp)
- Parsley (2 Tbsp., chopped)

Instructions

1. Combine all marinade ingredients and set aside.
2. Take shrimp in a bowl with 0.5 c. of the marinade. Let them sit for 30 minutes.
3. During rest time, set up your baking sheet with foil or parchment, and prepare your veggies. Chop them up and toss onto baking sheet, drizzling them with the olive oil and giving them a quick sprinkle of salt and pepper. Bake for roughly 20 minutes at 400F, then remove from oven. Take out all green beans and set to the side.
4. Place shrimp in one layer across the pan and bake for an additional 10 minutes until shrimp is done. Serve with veggies and shrimp in bowls, topped with 2 Tbsp. feta and a spoonful of yogurt marinade.

Mediterranean Mahi Mahi

Ingredients

- Basil (6 leaves, freshly chopped)
- Capers (4 Tbsp.)
- Garlic (2 cloves, chopped)
- Italian seasoning (pinch)
- Kalamata olives (25, chopped)
- Lemon juice (1 tsp)
- Mahi mahi (1 pound)
- Olive oil (2 Tbsp.)
- Onion (o.5, chopped)
- Parmesan cheese (3 Tbsp.)
- Diced tomatoes (15 oz. can)
- White wine (0.25 c.)

Instructions

1. Warm olive oil in a pan and then cook onions until translucent. Toss in garlic and seasoning and stir to mix well. Then, add in your can of tomatoes, wine, olives, lemon, and roughly half of the chopped basil. Drop heat down and toss in parmesan cheese. Cook until bubbling.
2. Put fish into a baking pan, then top with the sauce. Bake for 20 minutes at 425 F until fish is to temperature.

Chapter 7: Slow Cooker Meals

Slow Cooker Mediterranean Chicken

Ingredients

- Bay leaf (1)
- Capers (1 Tbsp.)
- Chicken broth (0.5 c.)

- Chicken thighs (2 pounds, bone and skin removed)
- Garlic (3 cloves, minced)
- Kalamata olives (1 c.)
- Olive oil (1 Tbsp.)
- Oregano (1 tsp)
- Roasted red pepper (1 c.)
- Rosemary (1 tsp, dried)
- Salt and pepper to taste
- Sweet onion (1, thinly sliced)
- Thyme (1 tsp, dried)
- Optional fresh lemon wedges to juice for serving

Instructions

1. Sauté the chicken in olive oil to brown on both sides, then remove it from the pan. Then, sauté the onions and garlic as well until beginning to soften, roughly 5 minutes.
2. Put chicken, onion, garlic, and all other ingredients into a slow cooker and leave it to cook for 4 hours on low. Season to taste.

Slow Cooker Vegetarian Mediterranean Stew

Ingredients

- Carrot (0.75 c., chopped)
- Chickpeas (15 oz. can)
- Crushed red pepper (0.5 tsp)
- Fire-roasted diced tomatoes (2 14-oz. cans)
- Garlic (4 cloves, minced)
- Ground pepper (0.25 tsp)
- Kale (8 c., chopped)

- Lemon juice (1 Tbsp.)
- Olive oil (3 Tbsp.)
- Onion (1, chopped)
- Oregano (1 tsp)
- Salt (0.75 tsp)
- Vegetable broth (3 c.)
- Basil leaves (garnish)
- Lemon wedges (garnish)

Instructions

1. Mix tomatoes, onion, carrot, broth, seasonings, and garlic into the slow cooker. Cook on low for 6 hours.
2. Take out 0.25 c. of the liquid in the slow cooker after 6 hours and transfer it to a bowl. Take out 2 Tbsp. of chickpeas and mash them with the liquid until nice and smooth.
3. Combine mash, kale, juice from lemon, and whole chickpeas. Cook for about 30 minutes, until kale is tender, then serve garnished with the basil leaves and lemon wedges.

Vegetarian Slow Cooker Quinoa

Ingredients

- Arugula (4 c.)
- Chickpeas (1 15.5 oz. can, rinsed and drained)
- Feta cheese (0.5 c)
- Garlic (2 cloves, minced)
- Kalamata olives (12, halved)
- Kosher salt (0.75 tsp)

- Lemon juice (2 tsp)
- Olive oil (2.25 Tbsp.)
- Oregano (2 Tbsp., fresh and coarsely chopped
- Quinoa (1.5 c., uncooked)
- Red onion (1 c., sliced)
- Roasted red pepper (0.5 c., drained and chopped)
- Vegetable stock (2.25 c.)

Instructions

1. Mix your broth with the onion, garlic, quinoa, chickpeas, and 1.5 tsp of olive oil. Sprinkle half of the salt atop it. Mix and cook on low until quinoa is done, roughly 3 or 4 hours.
2. Turn off the slow cooker and mix well. In a separate bowl, combine remaining olive oil, salt, and lemon juice together. Then, mix that into the slow cooker, along with the peppers.
3. Combine in the arugula and leave until the greens start to wilt. Serve, topping with feta, oregano, and olives.

Slow-Cooked Chicken and Chickpea Soup

Ingredients

- Artichoke hearts (14 oz. can, drained and chopped)
- Bay leaf (1)
- Cayenne (0.25 tsp)
- Chicken thighs (2 lbs., skins removed)
- Cumin (4 tsp)
- Diced tomatoes (1 15-ounce can)
- Dried chickpeas (1.5 c., allow to soak overnight)
- Garlic cloves (4, chopped)
- Olives (o.25 c., halved)
- Paprika (4 tsp)

- Pepper (0.25 tsp)
- Salt (0.5 tsp)
- Tomato paste (2 Tbsp.)
- Water (4 c.)
- Yellow onion (chopped)
- Parsley or cilantro (garnish)

Instructions

1. Drain your soaked chickpeas and place them into your slow cooker (large preferred). Mix in the water, onions and garlic, tomatoes (undrained), tomato paste, and all seasonings. Combine well, then add in the chicken.
2. Leave it to cook for 8 hours at low, or 4 at high.
3. Remove the chicken and allow it to cool on a cutting board. At the same time, remove the bay leaf, then add in the artichoke and olives. Season with additional salt if necessary to taste. Chop up chicken, removing the bones, and then mix it back into the soup. Serve the soup with the parsley or cilantro garnishing the top.

Slow Cooked Brisket

Ingredients

- Beef broth (0.5 c.)
- Brisket (3 lbs.)
- Cold water (0.25 c.)
- Fennel bulbs (2, cored, trimmed, and cut into wedges)
- Flour
- Italian seasoning (3 tsp)
- Italian seasoning diced tomatoes (14.5 oz. can)
- Lemon peel (1 tsp., fine shreds)
- Olives (0.5 c.)
- Parsley for garnish
- Pepper (pinch)
- Salt (pinch)

Instructions

1. Trim meat, then season with 1 tsp Italian seasoning. Put it in slow cooker with the cut-up fennel on top.
2. Mix together the tomatoes, broth, peel, olives, salt and pepper, and the last of the Italian seasoning.
3. Cook at low for 10 hours, or high for 5.
4. Take meat out of the cooker and reserve all juice. Arrange meat with veggies on a serving platter.
5. Remove fat from top of the juices.
6. Take 2 c. of juices in saucepan. Mix together water and flour, then combine it into the juice. Cook until gravy forms.
7. Serve meat topped with gravy and garnish with parsley.

Vegan Bean Soup with Spinach

Ingredients

- Vegetable broth (3 14-oz. cans)
- Tomato puree (15 oz. can)
- Great Northern or White beans (15 oz. can)
- White rice (0.5 c)
- Onion (0.5 c., chopped)
- Garlic (2 cloves, minced)
- Basil (1 tsp., dried)
- Pinch of salt
- Pinch of pepper
- Kale or spinach (8 c., chopped)

Instructions

1. Mix everything but leafy greens together in your slow cooker. Cook for 5 or 7 hours on low, or 2.5 hours on high.
2. Toss in leafy greens. Wait for them to wilt and serve.

Moroccan Lentil Soup

Ingredients

- Carrots (2 c., chopped)
- Cauliflower (3 c.)
- Cinnamon (0.25 tsp)
- Cumin (1 tsp)
- Diced tomato (28 oz.)
- Fresh cilantro (0.5 c.)
- Fresh spinach (4 c.)
- Garlic (4 cloves, minced)

- Ground coriander (1 tsp)
- Lemon juice (2 Tbsp.)
- Lentils (1.75 c.)
- Olive oil (2 tsp)
- Onion (2 c., chopped)
- Pepper (pinch)
- Tomato paste (2 Tbsp.)
- Turmeric (1 tsp)
- Vegetable broth (6 c.)
- Water (2 c.)

Instructions

1. Mix everything but spinach, cilantro, and lemon juice. Cook until lentils soften. This will be 4-5 hours if you use high heat, or 10 hours on low.
2. Mix spinach when just 30 minutes remains on cook time.
3. Just before serving, top with cilantro and lemon juice.

Chapter 8: Vegetarian and Vegan Meals

Vegetarian Greek Stuffed Mushrooms

Ingredients

- Cherry tomatoes (0.5 c., quartered)
- Feta cheese (0.33 c.)
- Garlic (1 clove, mixed)
- Ground pepper (0.5 tsp)
- Kalamata olives (2 Tbsp.)
- Olive oil (3 Tbsp.)
- Oregano (1 Tbsp., fresh and roughly chopped)
- Portobello mushrooms (4, cleaned with stems and gills taken out)
- Salt (0.25 tsp)
- Spinach (1 c., chopped)

Instructions

1. Begin by setting your oven. This recipe requires 400F for baking.
2. Mix together your salt and 0.25 tsp pepper, garlic, and 2 Tbsp. of oil, and use it to cover your mushrooms, inside and out.
3. Set the mushrooms onto your baking pan and allow it to cook for 10 minutes.
4. Mix together your remaining ingredients and combine well. Then, when the mushrooms are done, remove them from the oven and then fill them up with your filling.
5. Allow to cook for another 10 minutes.

Vegetarian Cheesy Artichoke and Spinach Stuffed Squash

Ingredients

- Artichoke Hearts (10 oz., frozen—thawed and chopped up)
- Baby spinach (5 oz.)
- Cream cheese (4 oz., softened)
- Parmesan cheese (0.5 c.)
- Pepper (pinch to taste)

- Red pepper and basil (for garnish)
- Salt (pinch to taste)
- Spaghetti squash (1, cut in half and cleaned out of seeds)
- Water (3 Tbsp.)

Instructions

1. Microwave your squash, flat side down, with 2 Tbsp. of your water uncovered for 10-15 minutes.
2. Mix together your spinach and water into a skillet until they begin to wilt. Then drain and reserve for later.
3. Preheat your oven set to broil with the rack at the upper 1/3 point.
4. Remove flesh from squash with a fork, then place the shells onto a sheet for the oven. Then stir in your artichoke, cheeses, and a pinch of salt and pepper to the squash flesh. Combine thoroughly, then split it between the two shells. Broil for 3 minutes and top with red pepper and basil to taste.

Vegan Mediterranean Buddha Bowl

Ingredients

For the chickpeas

- Chickpeas (1 can, rinsed, drained, and skinned)
- Olive oil (1 tsp)
- Pinch of salt and pepper
- Dried basil (0.25 tsp)
- Garlic powder (0.25 tsp)

For the quinoa

- Quinoa (0.5 c.)
- Water (1 c.)

For the salad

- Bell pepper (1, color of choice, seeded, stemmed, and chopped to bite-sized bits)
- Cucumbers (2, peeled and chopped)
- Grape tomatoes (1 c., halved)
- Hummus (0.5 c.)
- Kalamata olives (0.5 c.)
- Lettuce (2 c. – can sub in field greens, spinach, kale, or any other leafy greens)

Instructions

1. Set your oven up to prepare for baking. It should be at 0400F. Then, mix the ingredients for the chickpeas together, coating them evenly with the seasoning.
2. Put chickpeas in single layer and put them onto the baking sheet. Roast for 30 minutes with an occasional mixing and rotation of the pan to allow them all to cook evenly. Allow them to cool.
3. Start preparing the quinoa and water in a microwave-safe bowl. Combine the water and quinoa and microwave, covered, for 4 minutes. Then stir and microwave for 2 minutes longer. Give it one final stir and leave it to rest in the microwave for another minute or two.
4. Begin assembling your salad. Begin with the greens at the bottom, then top with tomatoes, cucumbers, bell pepper, olives, chickpeas, and then quinoa. Finally, top with a dollop of hummus to serve.

Vegan Mediterranean Pasta

Ingredients

- Artichokes (0.5 c.)
- Basil leaves (0.25 c., torn)
- Garlic cloves (2-3 to taste, minced)
- Grape tomatoes (2 c., halved)
- Kalamata olives (10, pitted)
- Olive oil (1 Tbsp.)

- Pasta (8 oz.)
- Red pepper (o.25 tsp.)
- Salt and pepper to taste
- Spinach (4 c.)
- Tomato paste (4 Tbsp.)
- Vegetable broth (1 c.)

Instructions

1. Prepare your pasta based on the instructions provided. Keep 1 c. of the water for later use and then set the pasta aside.
2. While preparing your pasta, take the time to warm a large skillet with oil. Then, sauté your garlic and red pepper for 30 seconds or so. Combine in the tomato paste and cook for another minute. At that point, mix in your tomatoes, your seasoning, your artichokes and olives, and your broth. Let it cook until tomatoes start to break down.
3. Mix in the pasta to the tomato mixture. Let it cook another 2 minutes and add reserved pasta water if too dry.
4. Add in spinach and basil and cook until wilted.
5. Remove from heat and serve.

Vegetarian Zucchini Lasagna Rolls

Ingredients

- Basil (2 Tbsp., fresh)
- Egg (1, lightly beaten)
- Frozen spinach (10-ounce package, thawed and dried)
- Garlic (1 clove)
- Marinara sauce (0.75 c.)
- Olive oil (2 tsp)
- Parmesan cheese (3 Tbsp.)
- Pinch each of salt and pepper
- Ricotta (1.33 c.)
- Shredded mozzarella cheese (8 Tbsp.)
- Zucchini (2, trimmed)

Instructions

1. Prepare two baking sheets with cooking spray. Then set the oven to 425F.
2. Cut up your zucchini into strips lengthwise into 1/8 inch thick pieces. A mandolin will make this easier.
3. Prepare zucchini coated in oil with salt and pepper, then set up a flat layer across the bottom of the prepared pan.
4. Bake zucchini for 10 minutes until it begins to soften.
5. Mix together 2 Tbsp. mozzarella and 1 Tbsp. of parmesan. Then, in another bowl, combine egg, ricotta, spinach, garlic, and the remainder of the cheese. Toss in a pinch of salt and pepper and mix well.

6. Set up an 8-inch square casserole dish with 0.25 c. marinara spread across the bottom.
7. Take your zucchini that has been softened and begin to roll it. To do this, you will need to put 1 Tbsp. of ricotta mix at the bottom of your strip, then roll. Put the seam down in the marinara-covered bottom. Do this for all pieces of zucchini.
8. Cover the rolls with the remainder of your marinara sauce and top with the cheese mix.
9. Bake until bubbling, roughly 20 minutes. Rest for 5 minutes and top with basil.

Vegetarian Breakfast Sandwich

Ingredients

- Sandwich thins (2)
- Olive oil (2 Tbsp. + 1 tsp)
- Rosemary (1 Tbsp. fresh, or 0.5 tsp dried)

- Eggs (2)
- Spinach leaves (1 c.)
- Tomato (0.5, sliced thinly)
- Feta (2 Tbsp.)
- Pinch of salt and pepper

Instructions

1. Warm oven to 375F. Separate your sandwich thins and coat with olive oil. Bake for 5 minutes until beginning to crisp up.
2. Warm skillet with last tsp of olive oil. Break eggs into pan and cook until whites are set. Then, break the yolks and flip to finish cooking.
3. Put bottoms of the bread onto serving plates. Then, top with spinach, the tomato, one egg each, followed by the feta. Sprinkle with salt and pepper, then top with remaining bread.

Vegan Breakfast Toast

Ingredients

- Bread of choice (verify that it is vegan—2 slices)
- Spice blend of choice
- Arugula (handful)
- Tomato (1, cut into rounds)
- Chopped olives (1 Tbsp.)
- Cucumber (0.5, cut into rounds)
- Hummus (0.25 c.)

Instructions

1. Toast up your bread. Then spread the hummus across, season it, and top with all toppings split between the pieces.

Vegetarian Shakshouka

Ingredients

- Chopped parsley (1 Tbsp.)
- Diced Tomatoes (15 oz. can)

- Eggs (4)
- Garlic (2 cloves)
- Olive oil (2 Tbsp.)
- Onion (1—sliced)
- Red bell peppers (2, sliced thinly)
- Salt and pepper to taste
- Spicy harissa (1 tsp)
- Sugar (1 tsp)

Instructions

1. Warm oil in a cast iron pan. Sauté your peppers and onions until they have begun to soften, giving them a stir every now and then to prevent sticking. Add in the garlic for another minute.
2. Put in tomatoes, sugar, and harissa, leaving it to simmer for the next 7 minutes.
3. Season it to taste. Then, add in small indentations into the mixture in the pan, cracking an egg in each indentation that you make. Cover up the pot and allow it to cook until egg whites are done.

Cover with parsley and serve with bread.

CPSIA information can be obtained
at www.ICGtesting.com
Printed in the USA
LVHW080549161220
674244LV00003B/103

9 789814 950749